Improving Guidance Programs

NORMAN C. GYSBERS

EARL J. MOORE

The University of Missouri

Prentice-Hall, Inc., Englewood Cliffs, New Jersey 07632

Library of Congress Cataloging in Publication Data

Gysbers, Norman C
Improving guidance programs.

Includes bibliographies and index.

1. Personal service in education. I. Moore, Earl J., joint author. II. Title.

LB1027.5.G94 371.4 80-21184
ISBN 0-13-452656-2

Editorial/production supervision
and interior design by Virginia Cavanagh Neri
Cover design by Maureen Olsen
Manufacturing buyer: Edmund W. Leone

Printed in the United States of America

10 9 8 7 6 5 4 3 2 1

PRENTICE-HALL INTERNATIONAL, INC., *LONDON*
PRENTICE-HALL OF AUSTRALIA PTY. LIMITED, *SYDNEY*
PRENTICE-HALL OF CANADA, LTD., *TORONTO*
PRENTICE-HALL OF INDIA PRIVATE LIMITED, *NEW DELHI*
PRENTICE-HALL OF JAPAN, INC., *TOKYO*
PRENTICE-HALL OF SOUTHEAST ASIA PTE. LTD., *SINGAPORE*
WHITEHALL BOOKS LIMITED, *WELLINGTON, NEW ZEALAND*

To Our Wives

Contents

3
SELECTING AND USING A DEVELOPMENTAL GUIDANCE PROGRAM MODEL 53

4
AN IMPROVED GUIDANCE PROGRAM: HOW TO GET THERE FROM WHERE ARE 103

5
CONDUCTING EVALUATION 149

APPENDIX 175

AUTHOR INDEX 203

SUBJECT INDEX 209

Preface

Since the early 1900s the structures and values of our society have undergone substantial changes. Industry, business, and labor have become highly complex and occupational specialization continues to increase. Concurrently, urbanization, a mobile population, the increasing depletion of natural resources, and the emergence of a variety of groups and movements asking for recognition have brought about extensive psychological, sociological, and economic changes. The challenges these and other changes represent to individuals and society have caused the educational community to continue to seek ways to improve and extend educational programs at all levels for all ages.

The individual and societal needs resulting from these changes are of particular concern for the guidance community because of the impact they have on the theory and practice of guidance in educational settings. It is clear from an analysis of these needs that a reconceptualization of guidance in education is necessary; from an ancillary, crisis-oriented only conception, to a comprehensive, developmental conception based on personal and societal needs organized programmatically around person-centered goals and activities designed to meet these needs.

Such a reconceptualization of guidance requires that the guidance program become an equal partner with the instructional program—with the concern for the intellectual development of individuals. Traditionally, guidance has not been conceptualized and implemented in this manner, because, as Aubrey (1973) suggested, guidance is still seen only as a supportive service that does not have a content base of its own. This same point was made by Sprinthall (1971) when he stated that there is little content in the practice of guidance and that guidance textbooks usually avoid discussion of a subject-matter base for guidance programs.

If guidance is to become an equal partner in education and meet the increasingly complex needs of individuals and society, it is our opinion that a content-base for guidance is required. The call for this is not new; many early guidance pioneers issued the same call. But the call was not loud enough during the early years so by the 1920s guidance had become essentially process or technique oriented. The need and the call continued to emerge occasionally thereafter however, but it wasn't until the late 1960s and early 1970s that it reemerged and became visible once more in the form of developmental guidance.

That is not to say that developmental guidance was not present before the late 1960s. What it does mean, however, is that by the late 1960s the need for attention to aspects of human development other than "the time-honored cognitive aspect of learning-subject matter mastery" (Cottingham 1974, p. 341) again had become apparent. Cottingham (1973) characterized these other aspects of human development as "personal adequacy learnings" (p. 342). Kehas (1973) pointed to this same need by stating that individuals should have opportunities "to develop intelligence about his self–his personal, unique, idiosyncratic, individual self" (p. 110). Similarly, Richardson and Baron (1975) suggested that two kinds of educational programs are needed today in education. The first kind they characterized as social learnings to be delivered through the instruction program primarily by teachers. The second kind they called personal learnings to be delivered through the counseling program primarily by guidance personnel.

It seems clear then, that the next step in the evolution of guidance is to establish guidance as a developmental program—a program that is an integral part of the educational process with a content base of its own. According to Kehas (1973), if we don't move in this direction we face status quo at best and extinction at worst.

The Introduction to this book which follows briefly traces the emergence of developmental guidance and its beginning implementation in the schools during the 1970s. Having an understanding of how this occurred is the first step in improving your school's guidance program. Based on this understanding we turn to the specifics of how guidance in schools can be improved. These specifics begin in Chapter 1. Chapter 1 focuses on the issues and concerns surrounding how to plan and organize for guidance program improvement. Chapter 2 discusses in detail the steps involved in finding out how well your current program is working and where improvement is needed. In Chapter 3, the importance of selecting and using a program model in the program improvement process is stressed. It is suggested that the model serve as an ideal against which you can compare the results of the current assessment of your present program. Based on this comparison, you can then adopt those aspects of the model which fit your situation, adapt other aspects of the model, or create new elements because of unique local needs. Chapter 4 describes in specific terms, how to structure, implement, and maintain your improved program or how to get there from where you are. Finally, Chapter 5 focuses on types of evaluation, reporting evaluation results, and using evaluation reports. In addition, emphasis is given to evaluation via a student competency reporting system.

The organization of the chapters may lead some readers to think that guidance program improvement activities follow one another in linear fashion. This is not the case however, since many of the activities described in Chapters 1 through 5 may be carried on concurrently. This is true particularly for the evaluation procedures described in Chapter 5. Many of these procedures are carried on from the beginning of the program improvement process throughout the life of the program. Some readers may misinterpret what has been written and think that guidance program improvement is a simple task requiring little staff time and few resources. This is not true. While substantial work can be completed during one year, at least two to three years with the necessary resources available to assure successful implementation is required.

REFERENCES

Aubrey, R.F., "Organizational Victimization of School Counselors," *School Counselor*, 20 (1973) 346-54.

Cottingham, H. F., "Psychological Education, the Guidance Function, and the School Counselor," *School Counselor*, 20 (1973) 340-45.

Kehas, C. D., "Guidance and the Process of Schooling: Curriculum and Career Education," *School Counselor*, 20 (1973) 109-15.

Richardson, H. D., and M. Baron, *Developmental Counseling in Education.* Boston: Houghton Mifflin Company, 1975.

Sprinthall, N. A., *Guidance for Human Growth*. New York: Van Nostrand Reinhold Company, 1971.

Introduction

During the late 1950s, the 1960s, and the early 1970s what we label as the counselor-clinical-services approach to guidance dominated professional theory, training, and practice. In schools those who adopted this approach emphasized the counselor and counseling. Administratively, guidance in the schools was part of pupil personnel services.

Guidance in the schools didn't begin that way, however. It began as vocational guidance with an emphasis on occupational selection and placement. But, in the 1920s, a more clinically oriented approach, which stressed counseling, began to emerge. While concern for occupational selection and placement was still present, a greater concern was expressed for personal adjustment. Thus the era of guidance for adjustment had begun.

Later, during the late 1920s and early 1930s, the beginnings of the services approach to organizing guidance in the schools appeared. Counseling continued to be featured as the dominant technique; only now it was called a service. By this time, too, the now traditional way of describing guidance as having three aspects–vocational, educational, and personal-social–had become well established. Vocational guidance, instead of being guidance, had become only one part of guidance. Finally, in the 1950s and early 1960s, guidance became a part of pupil personnel services.

THE CALL TO CHANGE

Beginning in the 1960s but particularly in the 1970s the concept of guidance for development emerged. The call came to reorient guidance from what had become an ancillary, crisis-oriented service to a comprehensive, developmental program. The call for reorientation came from diverse sources, including a renewed interest in vocational-career guidance and its theoretical base career development, concern about the efficacy of the prevailing approach to guidance in the schools, concern about accountability and evaluation, and from a renewed interest in developmental guidance.

Vocational-Career Guidance

The resurgence of interest in vocational-career guidance that began in the 1960s was aided, in part, by a series of national conferences on the topic. These conferences were funded through the Vocational Education Act of 1963 and later the Vocational Amendments of 1968. It is clear from Hoyt's (1974) account of these conferences that they contributed substantially to the renewed interest in the term *guidance* and its practice in the schools.

In addition, the resurgence of interest in vocational-career guidance was also aided by a number of career guidance projects begun in the 1960s. Among them was the Developmental Career Guidance Project, begun in 1964 in Detroit to provide career guidance for disadvantaged youth. It was one of the early developmental career guidance programs, one that accumulated sufficient evaluative data to support the further development of comprehensive guidance programming in schools (Leonard and Vriend 1975).

Concern about the Prevailing Approach

Paralleling the resurgence of interest in vocational-career guidance was a growing concern about the efficacy of the counselor-clinical-services approach in schools. Particular concern was expressed about an overemphasis on the one-to-one relationship model of counseling and the tendency for counselors to focus mainly on crises and problems.

> The traditional one-to-one relationship in counseling which we have cherished and perhaps overvalued will, of course, continue. But it is quite likely that the conception of the counselor as a roombound agent of behavior change must be critically reappraised. The counselor of the future will likely serve as a social catalyst, interacting in a two person relationship with the counselee part of the time, but also serving as a facilitator of the environmental and human conditions which are known to promote the counselee's total psychological development, including vocational development. (Borow 1965, p. 88)

This same issue was discussed from a slightly different perspective in an exchange between Brammer (1968) and Felix (1968). Brammer proposed the abandonment of the guidance model for counselors and the adoption of a counseling psychologist model in its place. Felix, in a reply to Brammer, sharply disagreed with Brammer's recommendation, pointing out that the counseling psychologist model wasn't valid for a school setting. Felix instead recommended an educational model for guidance. Similarly Aubrey (1969) recommended an educational model as opposed to a therapy model by pointing out that the therapy model was at odds or even frequently incongruent with educational objectives.

During the 1960s there were also expressions of concern about the potency of the guidance services concept and the need for more meaningful reconceptualizations for guidance if guidance were to reach higher levels of development (Roeber, Walz, and Smith 1969). This same theme was echoed by Sprinthall.

> It is probably not an understatement to say that the service concept has so dominated guidance and counseling that more basic and significant questions are not even acknowledged, let alone answered. Instead, the counselor assumes a service orientation that limits and defines his role to minor administrative procedures.[1]

[1] N.A. Sprinthall, *Guidance for Human Growth,* ©1971 by Litton Educational Publishing, Inc. Reprinted by permission of Van Nostrand Reinhold Company.

The Accountability Movement

The call for change in guidance was reinforced by the accountability movement in education, which had begun during the 1960s. As education was being held accountable for its outcomes, so too was guidance. It was apparent that it would be necessary for counselors to state guidance goals and objectives in measurable outcome terms and show how these goals and objectives were related to the general goals of education. Dickinson (1969–70) made this point when he stated that "counselors must turn their attention to setting specific goals if we are to remain a major force in education. This is going to be a difficult task, and we must begin now." (p. 16)

Wellman and Twiford (1961) also stressed this point when they stated that the one appropriate measure of the value of a guidance program was its impact on students. Later in the 1960s personnel of the National Study of Guidance under the direction of Wellman (1968) developed a systems model for evaluation. A taxonomy of guidance objectives classified in three domains of educational, vocational, and social development accompanied the model. Wellman's model and its companion taxonomy of objectives served as a basis for a number of evaluation models that began appearing in the late 1960s and early 1970s. A *Process Guide for the Development of Objectives,* originally published by the California State Department of Education in 1970 and later published by the California Personnel and Guidance Association (Sullivan and O'Hare 1971), is an example of one such model.

The accountability movement, with its focus on measurable outcomes, presented a real problem for counselors, however. The traditional service approach emphasized *techniques* of guidance rather than *purposes* of guidance (Sprinthall 1971). As a result, counselors were known for the techniques they used, not for the outcomes these techniques produced in individuals. This perspective was supported by counselor education programs, since the focus on such programs tended to be mainly on counselors and techniques and not as much on guidance program outcomes. Apparently, it was assumed that guidance techniques, particularly individual counseling, were difficult to learn and, therefore, a majority of training time needed to be devoted to them. Learning how to develop and manage a guidance program with an emphasis on measurable outcomes, it was felt, needed not be stressed as much, since such things could be learned on the job (Gysbers 1969a).

Developmental Guidance

Finally, in the 1960s, the term *developmental guidance* was heard with increasing frequency. Mathewson (1962), in discussing future trends for guidance, suggested that although adjustive guidance was popular, a long-term movement toward developmental forms of guidance would probably prevail.

> In spite of present tendencies, a long-term movement toward educative and developmental forms of guidance in schools may yet prevail for these reasons: the need to develop all human potentialities, the persistence and power of human individuality, the effects of dynamic educative experience, the necessity for educational adaptability, the comparative costs, and the urge to preserve human freedom.[2]

Similarly, Zaccaria (1966) stressed the importance of and need for developmental guidance. He pointed out that developmental guidance was a concept in transition, that it was in tune with the times but still largely untried in practice.

DEVELOPMENTAL PROGRAMS EMERGE

In the early 1970s the accountability movement intensified. It was joined by increasing interest in career development theory, research, and practice and its educational manifestations, career guidance and career education. Other educational movements such as psychological education, moral education, and process education also emerged. In addition, interest in the development of

[2] R.H. Mathewson, *Guidance Policy and Practice*, 3rd ed., ©1962, p. 375. Reprinted by permission of Harper & Row, Publishers, Inc.

comprehensive systematic approaches to guidance program development and management continued to increase. The convergence of these movements in the early 1970s served as a stimulus to continue the task of defining guidance developmentally in measurable individual outcome terms—as a program in its own right rather than as services ancillary to other programs.

By 1970 a substantial amount of preliminary work had been done in developing basic ideas, vocabulary, and constructs to define guidance in comprehensive-developmental-outcome terms. As early as 1961 Glanz identified and described four basic models for organizing guidance because of his concern about the lack of discernible patterns for implementing guidance in the schools. Tiedeman and Field (1962) issued a call to make guidance an integral part of the educational process. They also stressed the need for a developmental, liberating perspective of guidance. Zaccaria (1965) stressed the need to examine developmental tasks as a basis for determining the goals of guidance. Shaw and Tuel (1966) developed a model for a guidance program designed to serve all students. At the elementary level Dinkmeyer (1966) emphasized the need for developmental counseling by describing pertinent child development research that supported a developmental perspective.

Paralleling the preliminary work on ideas, vocabulary, and constructs was the application of systems thinking to guidance. Based on a nationwide survey of vocational guidance in 1968, a systems model for vocational guidance was developed at the Center for Vocational and Technical Education in Columbus, Ohio. The model focused on student behavioral objectives, alternative activities, program evaluation, and implementation strategies (Campbell, Dworkin, Jackson, Hoeltzel, Parsons, and Lacey 1971). Ryan (1969), Thoresen (1969), and Hosford and Ryan (1970) also proposed the use of systems theory and systems techniques for the development and improvement of comprehensive guidance programs.

On the West Coast McDaniel (1970) proposed a model for guidance called Youth Guidance Systems. It was organized around goals, objectives, programs, implementation plans, and designs for evaluation. The primary student outcome in this model was considered to be decision making. Closely related to this model was the Comprehensive Career Guidance System (CCGS) developed by personnel at the American Institutes for Research (Jones, Nelson,

Ganschow, and Hamilton 1971; Jones, Hamilton, Ganschow, Helliwell, and Wolff 1972). The CCGS was designed to plan, implement, and evaluate guidance programs systematically. Systems thinking also undergirded Ryan and Zeran's (1972) approach to the organization and administration of guidance services. They stressed the need for a systems approach to guidance in order to insure the development and implementation of an accountable program. Finally, a systematic approach to guidance was advocated in the PLAN (Program of Learning in Accordance with Needs) System of Individualized Education (Dunn 1972). Guidance was seen as a major component of PLAN and was treated as an integral part of the regular instructional program.

The task of defining guidance in comprehensive-developmental-outcome terms received substantial support from these approaches that applied systems thinking to guidance. Additional support was provided by the development in a number of states in the early 1970s of state guides for integrating career development into the school curriculum. One such guide was developed in August 1970 by the state of Wisconsin (Drier 1971), closely followed by the development of the California Model for Career Development in the summer of 1971 (California State Department of Education 1971).

The idea of implementing career development through the curriculum did not, of course, originate with these models. As early as 1914 Davis had outlined such a curriculum. Of more immediate interest, however, is the work of Tennyson, Soldahl, and Mueller (1965) titled *The Teacher's Role in Career Development* and the Airlie House Conference in May 1966 on the topic "Implementing Career Development Theory and Research through the Curriculum," sponsored by the National Vocational Guidance Association (Ashcraft 1966). Later in the 1960s and early 1970s came the work of such theorists and practitioners as Gysbers (1969b), Herr (1969), Hansen (1970), and Tennyson and Hansen (1971), all of whom spoke to the need to integrate career development concepts into the curriculum. Through these efforts and others like them, career development concepts began to be translated into individual outcomes and the resulting goals and objectives arranged sequentially, K–12.

Concurrent with these efforts, a national effort was begun to assist the states in developing and implementing state models or guides for career guidance, counseling, and placement. On July 1,

1971, the University of Missouri–Columbia was awarded a U.S. Office of Education grant to assist each state, the District of Columbia, and Puerto Rico in developing models or guides for implementing career guidance, counseling, and placement programs in local schools. This project was the next step in a program of work begun as a result of a previous project at the university, a project that conducted a national conference on career guidance, counseling, and placement in October 1969 and regional conferences across the country during the spring of 1970. All fifty states, the District of Columbia, and Puerto Rico were involved in the 1971 project, and by the time the project ended in 1974, forty-four states had developed some type of guide or model for career guidance, counseling, and placement. As a part of the assistance provided to the states, project staff conducted a national conference in January 1972 and developed a manual (Gysbers and Moore 1974) to be used by the states as they developed their own guides.

By the early 1970s it was clear that the movement toward developing and implementing comprehensive, developmental guidance programs was well under way. Influenced by career development theory and research, the accountability-evaluation movement, and systems thinking, the earlier promise of guidance for development began to take on form and substance. Career development theory and research offered the content and objectives (Herr and Cramer 1972; Walz, Smith, and Benjamin 1974; Tennyson, Hansen, Klaurens, and Antholz 1975); the emphasis on accountability-evaluation provided the impetus, knowledge, and methods to plan, structure, implement, and judge guidance programs (O'Hare and Lasser 1971; Mease and Benson 1973; Wellman and Moore 1975); and systems thinking provided a way to systematically organize evaluation (Ryan 1969; Thoresen 1969; Hosford and Ryan 1970; McDaniel 1970; Jones, Nelson, Ganschow and Hamilton 1971; Jones, Hamilton, Ganschow, Helliwell and Wolff 1972; Ryan and Zeran 1972).

DEVELOPMENTAL PROGRAMMING CONTINUES TO EXPAND

As the 1970s continued to unfold, professional literature devoted to the why and how of developing and implementing systematic accountable guidance programs continued to be written. Humes (1972) urged consideration of applying the planning, programming, budgeting system (PPBS) to guidance programs. Hays (1972) stressed the need for guidance programs to be accountable. Pulvino and Sanborn (1972) underlined the same point and then described a communications system for planning and carrying out guidance and counseling activities. A more cautious call to accountability through behavioral objectives was issued by Koch (1974). He outlined possible negative side effects and then listed conditions through which behavioral objective writing for guidance could come to fruition. In a similar manner Gubser (1974) stressed the point that for counselor accountability to take place the school system must also be accountable. One cannot be accountable independent of the other.

As the movement toward planning and implementing systematic developmental and accountable guidance programs in the early 1970s became more sophisticated, theoretical models began to be translated into practical, workable models to be implemented in the schools. One vehicle used in this translation process was an expanded conception of career guidance. An example effort in this regard began in 1972 in Mesa, Arizona (McKinnon and Jones 1975). The guidance staff in Mesa felt the need to reorient their guidance program to make it more accountable. The vehicle they chose to do this was a comprehensive career guidance program that included needs assessment, goals and objectives development, and related guidance activities. In cooperation with the American Institutes for Research competency-based training packages were written to train staff in program development and implementation methods and procedures.

A similar example effort was begun at the Grossmont Union High School District in the state of California in 1974 (Jacobson and Mitchell 1975). Guidance personnel in the district chose the *California Model for Career Development* (California State Department of Education 1971) to supply the content of the program and then proceeded to lay out a systematic, developmental career guidance program. Another example occurred in Georgia when the Georgia State Department of Education initiated a project funded by the U.S. Office of Education to coordinate the efforts of several Georgia school systems in planning and implementing comprehensive career guidance programs. The goal of the project was to develop a career guidance system based on student needs, which focused on a team approach and curriculum-based strategies (Dagley 1974).

On July 1, 1974, the American Institutes for Research began work on bringing together program planning efforts previously undertaken by the Pupil Personnel Division of the California State Department of Education and their own Youth Development Research Program in Mesa, Arizona, and elsewhere (Jones, Helliwell, and Ganschow 1975). This resulted in the development of twelve competency-based staff development modules on Developing Comprehensive Career Guidance Programs K–12. As a part of the project, the modules were field-tested in two school districts in California in the summer of 1975 and in a preservice class of guidance and counseling majors at the University of Missouri–Columbia in the fall of 1975. A final report of this project was issued by the American Institutes for Research in January 1976 (Dayton 1976). Later Jones, Dayton, and Gelatt (1977) used the twelve modules as a point of departure to suggest a systematic approach in planning and evaluating human service programs.

The work that began in the early 1970s on guidance program models was continued and expanded as the 1970s unfolded. In May 1975 a special issue of the *Personnel and Guidance Journal* on "Career Development: Guidance and Education," edited by Hansen and Gysbers (1975) was published. In it a number of articles described program models and examples of programs in operation. In 1976 the American College Testing Program published a programmatic model for guidance. It was titled *River City High School Guidance Services: A Conceptual Model* (American College Testing Program 1976).

The systems approach to program development, first emphasized in the late 1960s, continued to be an important emphasis in the later part of the 1970s. Ewens, Dobson, and Seals (1976) described and discussed a systems approach to career guidance K–12 and beyond. In a similar fashion, although using the traditional services model, Ryan (1978) presented a systems approach to the organization and administration of guidance services.

By the later part of the 1970s an increasing number of articles, monographs, and books were being published on various aspects of comprehensive guidance programming. Brown (1977) discussed the organization and evaluation of elementary school guidance services using the three-C's approach of counseling, consulting, and coordinating. Upton, Lowrey, Mitchell, Varenhorst, and Benvenuti (1978) described procedures for developing a career guidance curriculum and presented leadership strategies to teach the procedures to those who would implement the curriculum. Ballast and Shoemaker (1978) outlined and described a step-by-step approach to developing a comprchensive K–12 guidance program. Campbell, Rodebaugh, and Shaltry (1978) edited a handbook that presented numerous examples of career guidance programs, practices, and models. Herr and Cramer (1979) described and discussed a systematic planning approach for career guidance, delineating goals, objectives, and activities for elementary, junior, and high schools as well as for higher and adult education. And finally, Hilton (1979) provided a conceptual framework for career guidance in the secondary school.

In an article in the handbook by Campbell, Rodebaugh, and Shaltry, Gysbers (1978) listed and described a number of systematic approaches to comprehensive guidance programming, including the *Career Planning Support System* (Campbell 1977) and the *Cooperative Rural Guidance System* (Drier 1976), both developed at the National Center for Research in Vocational Education, Columbus, Ohio. Another, similar approach developed during the later 1970s was the *Programmatic Approach to Guidance Excellence: PAGE 2* (Peterson and Treichel 1978). Finally, articles by Mitchell (1978) and Mitchell and Gysbers (1978) described the need for comprehensive guidance programs and provided recommendations for how to develop and implement such programs, and a publication by Halasz-Salster and Peterson (1979) presented descriptions of different guidance planning models.

The later half of the 1970s also witnessed increasing legislative activity to mandate comprehensive, developmental guidance programming in schools. For example, developmental elementary school guidance was identified as a critical area of education in the state of Oklahoma. The 1978 state legislature appropriated $1.72 million to be distributed to schools on the basis of $5,000 per program (Fisher 1978). Grants are made to eligible schools provided the schools meet these minimum criteria: statement of philosphy; needs assessment; program goals; objectives; activities; evaluation of objectives; program evaluation, review, and modification; personnel; student ratio; physical facilities; materials and equipment; budget; cross-level program articulation, and professional growth and development. Similar mandating efforts were undertaken by a task force of guidance professionals in California in 1977 (Hooper 1977). In addition, on September 28, 1979, a bill, H.R. 5477, titled "Elementary School Guidance and Counseling Incentive Act of 1979," was introduced in the U.S. House of Representatives. The purpose of this bill was to "assure the accessibility of developmental guidance and counseling to all children of elementary school age by providing funds for comprehensive elementary school guidance and counseling programs."

At the same time that attention was being given to state and federal legislation and mandation, some states were developing planning models for guidance. For example, personnel at Marshall University undertook a project to develop a planning model and a state plan for improving comprehensive systems of career guidance in West Virginia. The plan includes community-based guidance services, upgrading guidance services personnel, and evaluation of guidance services programs for students, out-of-school youth, and adults (West Virginia Department of Education, 1979).

Thus, as the 1970s drew to a close, it was clear that traditional approaches to guidance and counseling in the nation's schools were being reexamined and that new patterns were being recommended (Herr 1979). The counselor-clinical-services model of the 1960s was being encompassed gradually by comprehensive, developmental programming. And, more often than not, career development was being used as the point of departure for such programming. Legislative language at the federal and state level confirmed this, as did the language of many state and local school plans for guidance (Drier and Herr 1978).

REFERENCES

American College Testing Program, *River City High School Guidance Services: A Conceptual Model.* Iowa City, Iowa; American College Testing, 1976.

Ashcraft, K. B., *A Report of the Invitational Conference in Implementing Career Development Theory.* Washington, D.C.: National Vocational Guidance Association, 1966.

Aubrey, R. F., "Misapplication of Therapy Models to School Counseling," *Personnel and Guidance Journal,* 48 (1969), 273-278.

Ballast, D. L. and R. L. Shoemaker, *Guidance Program Development.* Springfield, Ill.: Charles C Thomas, Publisher, 1978.

Borow, H., "Research in Vocational Development: Implications for the Vocational Aspects of Counselor Education," in *Vocational Aspects of Counselor Education,* ed. C. McDaniels. Washington, D.C.: George Washington University, 1966. A report of a conference at Airlie House, George Washington University.

Brammer, L. M., "The Counselor Is a Psychologist," *Personnel and Guidance Journal,* 47 (1968) 4-9.

Brown, J. A., *Organizing and Evaluating Elementary School Guidance Services: Why, What, and How.* Monterey, Calif.: Brooks/Cole Publishing Co., 1977.

California State Department of Education, *Career Guidance: A California Model for Career Development K-Adult.* December 1971.

Campbell, R. E., *The Career Planning Support System.* Columbus, Ohio: National Center for Research in Vocational Education, 1977.

Campbell, R. E., E. P. Dworkin, D. P. Jackson, K. E. Hoeltzel, G. E. Parsons, and D. W. Lacey, *The Systems Approach: An Emerging Behavioral Model for Career Guidance.* Columbus, Ohio: Center for Vocational and Technical Education, 1971.

Campbell, R. E., H. D. Rodebaugh, and P. E. Shaltry, *Building Comprehensive Career Guidance Programs for Secondary Schools.* Columbus, Ohio: National Center for Research in Vocational Education, 1978.

Dagley, J. C., *Georgia Career Guidance Project Newsletter,* Athens: University of Georgia, December 1974.

Davis, J. B., *Vocational and Moral Guidance.* Boston: Ginn & Company, 1914.

Dayton, C. A., *A Validated Program Development Model and Staff Development Prototype for Comprehensive Career Guidance, Counseling, Placement and Follow-up.* Palo Alto, Calif.: American Institutes for Research Final Report, Grant no. OEG-0-74-1721, January 1976.

Dickinson, D. J., "Improving Guidance with Behavioral Objectives," *CPGA Journal,* II (1969), 12-17.

Dinkmeyer, D., "Developmental Counseling in the Elementary School," *Personnel and Guidance Journal,* 45, (1966) 262-66.

Drier, H. N., *Cooperative Rural Guidance System.* Columbus, Ohio: National Center for Research in Vocational Education, 1976.

Drier, H. N., ed. *Guide to the Integration of Career Development into Local Curriculum–Grades K-12.* Madison, Wis.: Wisconsin Department of Public Instruction, December 1971.

Drier, H. N. and E. L. Herr, *Solving the Guidance Legislative Puzzle.* Washington, D.C.: American Personnel and Guidance Association and American Guidance Association, 1978.

Dunn, J. A., *The Guidance Program in the Plan System of Individualized Education.* Palo Alto, Calif.: American Institutes for Research, 1972.

Ewens, W. P., J. S. Dobson, and J. M. Seals, *Career Guidance: A Systems Approach.* Dubuque, Iowa: Kendall-Hunt Publishing Company, 1976.

Felix, J. L., "Who Decided That?" *Personnel and Guidance Journal,* 47 (1968), 9-11.

Fisher, L., Letter to the superintendents of schools in Oklahoma, May 15, 1978.

Glanz, E. C., "Emerging Concepts and Patterns of Guidance in American Education," *Personnel and Guidance Journal,* 40 (1961) 259-65.

Gubser, M. M., "Performance-based counseling: Accountability or Liability," *School Counselor,* 21 (1974) 296–302.

Gysbers, N. C., "Educating School Counselors," *Contemporary Education,* November supplement, 1969a.

———, "Comprehensive Career Guidance Programs," in *Building Comprehensive Career Guidance Programs for Secondary Schools,* eds. R. E. Campbell, H. D. Rodebaugh, and P. E. Shaltry. Columbus, Ohio: National Center for Research in Vocational Education, 1978.

———, "Elements of a Model for Promoting Career Development in Elementary and Junior High School." Paper presented at the National Conference on Exemplary Programs and Projects, 1968 Amendments to the Vocational Education Act (ED045860), Atlanta, Georgia, March 1969b.

Gysbers, N. C. and E. J. Moore., eds. *Career Guidance Counseling and Placement: Elements of an Illustrative Program Guide.* Columbia, Mo.: University of Missouri, 1974.

Halasz-Salster, I. and M. Peterson, *Planning Comprehensive Career Guidance Programs: A Catalog of Alternatives.* Columbus, Ohio: National Center for Research in Vocational Education, 1979.

Hansen, L. S., *Career Guidance Practices in School and Community.* Washington, D.C.: National Vocational Guidance Association, 1970.

Hansen, L. S. and N. C. Gysbers, eds. "Career Development: Guidance and Education," *Personnel and Guidance Journal*, 53 (special issue, 1975).

Hays, D. G., "Counselor–What Are You Worth?" *School Counselor,* 19 (1972) 309-12.

Herr, E. L., *Guidance and Counseling in the Schools: The Past, Present, and Future.* Washington, D.C.: American Personnel and Guidance Association, 1979.

———, "Unifying an Entire System of Education around a Career Development Theme," Paper presented at the National Conference on Exemplary Programs and Projects, 1968 Amendments to the Vocational Education Act (ED045860), Atlanta, Georgia, March 1969.

Herr, E. L. and S. H. Cramer, *Career guidance through the Life Span.* Boston: Little, Brown & Company, 1979.

———, *Vocational Guidance and Career Development in the Schools: Toward a Systems Approach.* Boston: Houghton Mifflin Company, 1972.

Hilton, T. L., *Confronting the Future: A Conceptual Framework for Secondary School Career Guidance.* New York: College Entrance Examination Board, 1979.

Hooper, P., Memo dated September 14, 1977.

Hosford, R. E. and A. T. Ryan, "Systems Design in the Development of Counseling and Guidance Programs," *Personnel and Guidance Journal,* 49 (1970) 221-30.

Hoyt, K. B., "Professional Preparation for Vocational Guidance," in *Vocational Guidance and Human Development,* ed. E. Herr. Boston: Houghton Mifflin 1974.

Humes, C. W., Jr., "Program Budgeting in Guidance," *School Counselor,* 19 (1972) 313-18.

Jacobson, T. J., and A. M. Mitchell, *Master Plan for Career Guidance and Counseling.* Final Report, Pupil Personnel Services, Grossmont Union High School District, June 30, 1975.

Jones, G. B., C. Dayton, and H. B. Gelatt, *New Methods for Delivering Human Services.* New York: Human Services Press, 1977.

Jones, G. B., J. A. Hamilton, L. H. Ganschow, C. B. Helliwell, and J. M. Wolff, *Planning, Developing, and Field Testing Career Guidance Programs: A Manual and Report.* Palo Alto, Calif.: American Institutes for Research, 1972.

Jones, G. B., C. B. Helliwell, and L. H. Ganschow, "A Planning Model for Career Guidance." *Vocational Guidance Quarterly,* 23 (1975), 220-26.

Jones, G. B., D. E. Nelson, L. H. Ganschow, and J. A. Hamilton, *Development and Evaluation of a Comprehensive Career Guidance System.* Palo Alto, Calif.: American Institutes for Research, 1971.

Koch, J. H., "Riding the Behavioral Objective Bandwagon," *School Counselor,* 21 (1974), 196-202.

Leonard, G. E., and T. J. Vriend, "Update: The Developmental Career Guidance Project," *Personnel and Guidance Journal,* 53 (1975), 668-71.

McDaniel, H. B., *Youth Guidance Systems.* Palo Alto, Calif.: College Entrance Examination Board, 1970.

McKinnon, B. E., and G. B. Jones, "Field Testing a Comprehensive Career Guidance Program, K-12," *Personnel and Guidance Journal,* 53 (1975), 663-67.

Mathewson, R. H., *Guidance Policy and Practice* (3rd ed.). New York: Harper & Row, Publishers, Inc., 1962.

Mease, W. P., and L. L. Benson, *Outcome Management Applied to Pupil Personnel Services.* St. Paul, Minnesota Department of Education, 1973.

Mitchell, A. M., "The Design, Development and Evaluation of Systematic Guidance Programs," *New Imperatives for Guidance,* ed. in G. Walz and L. Benjamin. Ann Arbor, Mich.: ERIC Counseling and Personnel Services Clearinghouse, 1978.

Mitchell, A. M., and N. C. Gysbers, "Comprehensive School Guidance Programs," in *The Status of Guidance and Counseling in the Nation's Schools.* Washington, D.C.: American Personnel and Guidance Association, 1978.

O'Hare, R. W. and B. Lasser, *Evaluating Pupil Personnel Programs.* Fullerton, Calif.: California Personnel and Guidance Association, 1971.

Peterson, M., and J. Treichel, *Programmatic Approach to Guidance Excellence, PAGE 2* (Rev. ed.). McComb, Ill.: Curriculum Publishing Clearinghouse, Western Illinois University, 1978.

Pulvino, C. J., and M. P. Sanborn, "Feedback and Accountability," *Personnel and Guidance Journal,* 51 (1972) 15-20.

Roeber, E. C., G. R. Walz, and G. E. Smith, *A Strategy for Guidance.* New York: Macmillan, Inc., 1969.

Ryan, T. A., *Guidance Services.* Danville, Ill.: The Interstate Printers & Publishers, Inc., 1978.

——, "Systems Techniques for Programs of Counseling and Counselor Education," *Educational Technology,* 9 (1969), 7-17.

Ryan, T. A., and F. R. Zeran, *Organization and Administration of Guidance Services.* Danville, Ill.: The Interstate Printers & Publishers, Inc., 1972.

Shaw, M. C., and J. K. Tuel, "A Focus for Public School Guidance Programs: A Model and Proposal," *Personnel and Guidance Journal,* 44 (1966), 824-30.

Sprinthall, N. A., *Guidance for Human Growth.* New York: Van Nostrand Reinhold Company, 1971.

Sullivan, H. J., and R. W. O'Hare, *A Process Guide for the Development of Objectives.* Fullerton, Calif.: California Personnel and Guidance Association, 1971.

Tennyson, W. W., and L. S. Hansen, "Guidance through the Curriculum," in *The Encyclopedia of Education* No. 4, ed. L. C. Deighton. New York: Macmillan, Inc., 1971, pp 248-54.

Tennyson, W. W., T. A. Soldahl, and C. Mueller, *The Teacher's Role in Career Development.* Washington, D.C.: National Vocational Guidance Association, 1965.

Tennyson, W. W., L. S. Hansen, M. K. Klaurens, and M. B. Anholz, *Educating for Career Development.* St. Paul, Minn.: Minnesota Department of Education, 1975.

Thoresen, C. E., "The Systems Approach and Counselor Education: Basic Features and Implications," *Counselor Education and Supervision,* 9 (1969), 3-17.

Tiedeman, D. V., and F. C. Field, "Guidance: The Science of Purposeful Action Applied through Education," *Harvard Educational Review,* 32 (1962), 483-501.

Upton, A., B. Lowrey, A. M. Mitchell, B. Varenhorst, and J. Benvenuit, *A Planning Model for Developing Career Guidance Curriculum.* Fullerton. Calif.: California Personnel and Guidance Association, 1978.

Walz, G. R., R. L. Smith, and L. Benjamin, *A Comprehensive View of Career Development.* Washington, D.C.: American Personnel and Guidance Association, 1974.

Wellman, F. E., *Contractor's Report, Phase I, National Study of Guidance.* Contract OEG 3-6-001147-1147. Washington, D.C.: U.S. Department of Health, Education and Welfare, Office of Education, 1968.

Wellman, F. E., and E. J. Moore, *Pupil Personnel Services: A Handbook for Program Development and Evaluation.* Columbia, Mo.: Missouri Evaluation Projects, 1975.

Wellman, F. E., and D. D. Twiford, *Guidance, Counseling and Testing: Program Evaluation.* Washington, D.C.: U.S. Department of Health, Education and Welfare, 1961.

West Virginia Department of Education, *A Planning Model for the Formulation of State and Local Career and Vocational Guidance Plans: Part IV, Program Narrative,* 1979.

Zaccaria, J. S., Developmental Guidance: A Concept in Transition," *School Counselor,* 13 (1966), 226-29.

———, "Developmental Tasks: Implications for the Goals of Guidance," *Personnel and Guidance Journal,* 44 (1965), 372-75.

Chapter 1

Getting Organized

> As public expectations increase for individualizing the educational process, as some options open for education beyond high school and as the job markets shift with the ebb and flow of social and technological change, the challenge and responsibilities resting upon the country's school counselors becomes steadily more demanding.[1]

School counselors are today expected to be involved in a greater variety of guidance and counseling activities than ever before. They are expected to work in the curriculum; conduct placement, follow-up, and follow-through activities; and do community outreach. In addition, they are expected to continue such guidance functions as crisis counseling and teacher and parent consultation as well as testing, scheduling, and other administrative-clerical duties. Most school counselors want to respond to these new expectations but often find that the press of their existing duties interferes with or actually prevents them from doing so.

At the same time, many school counselors find that current organizational patterns of guidance are based on the ancillary services concept. And as a result, they find themselves in a mainly supportive, remedial role, a role that is not seen as mainstream by most people. And what is worse, this concept reinforces the practice of having counselors do many quasi-guidance tasks because

[1] Reprinted with permission from *The College Board News*, September 1977. © 1977 by College Entrance Examination Board, New York.

such tasks can be justified as being of service to someone. The following list is typical of such tasks.

Counselors register and schedule all new students.

Counselors are responsible for giving ability and achievement tests.

Counselors talk to new students concerning school rules.

Students go to the counselor to get their schedules changed.

Counselors are responsible for signing excuses for students who are tardy or absent.

Counselors teach classes when teachers are absent.

Counselors do senior grade checks.

Counselors are assigned lunchroom duty.

Counselors arrange class schedules for students.

Students are sent to the counselor for disciplinary action.

Counselors send students home who are not appropriately dressed.

Counselors compute grade-point averages.

Counselors fill out student reports and records.

Counselors are in charge of student records.

Counselors supervise study halls.

Counselors assist with duties in the principal's office.

Given this situation, the challenge that counselors face is how to make the transition from the ancillary service concept of guidance to that of a comprehensive, developmental program, a program that is an equal partner with other programs in education. Making the transition is a complex and difficult task because it means carrying out duties provided by the current organizational plan at the same time as planning and trying out new duties derived from a new organizational plan. It can be done, but it is difficult, time consuming, and often frustrating.

The initial stimulus to modify guidance may come from counselors, or it may come from parents, students, school administration, the school board, or community organizations. No matter where the initial stimulus comes from, however, the total guidance staff (preferably K–12) must be involved in responding to it. Time and privacy are required for a staff to evolve their response. When all the pros and cons of change have been discussed fully, consensus must be reached on how to respond. The staff may decide to stay as is. Or, the decision may be made to change. If the decision is made to change, then it is imperative that a majority of the staff agree to be involved fully in whatever it takes to improve their guidance program, to make it comprehensive and developmental.

IDENTIFY THE STEPS, PROCESSES, AND MATERIALS INVOLVED

Because modifying an existing program or planning and implementing a new program is complex, one of the first steps is to identify the tasks involved, the change processes to be used, and any materials or resources that may be required. This is an important step to complete in the beginning because it provides those involved with a map of the components and the sequence to follow. In addition, it can indicate the relationship among components.

A number of writers have addressed this issue and have outlined and described what may be involved in the program planning and implementing process. Ballast and Shoemaker (1978, p. 7) defined program development as a systematic process that includes the following components and sequence: identifying guidance department needs, generating support for program development, establishing departmental leadership, preparing the proposal for leadership development, involving key decision makers, identifying current services and activities, developing tools for the assessment of student needs, administering needs surveys, tabulating the results of needs surveys, identifying priority needs survey items, interpreting the results of the needs surveys, developing student outcome statements, determining counselor activities designed to reach student outcomes and integrated with current activities and services, identifying timelines and materials, developing a guidance calendar including individual counselor responsibilities, and organizing the guidance program handbook.

Mitchell and Gysbers (1978) described a comprehensive guidance program model having four major phases: planning, designing, implementing, and evaluating. Each of the phases contains specific tasks to be completed. The phases and tasks they described are as follows:

Planning

1. Statement of values
2. Selection of a curriculum model
3. Selection of program goals
4. Determination of desired student outcomes
5. Assessment of current program
6. Establishment of priorities

Designing

1. Development of program objectives
2. Selection of program strategies
3. Assignment of program components
4. Analysis of staff competencies
5. Provision of staff development

Implementing

1. Administration of assessment instruments
2. Installation of program
3. Identification based on evaluation data

Evaluating

1. Formulation of the questions to be answered by the evaluation
2. Selection of evaluation design
3. Selection of assessment instruments
4. Development of procedures for data collection
5. Establishment of a monitoring system
6. Performance of data reduction, summary, and analysis tasks
7. Preparation of reports

Augmenting the work of these writers, a number of other writers and organizations have developed guidance program planning and implementation models. Ryan (1978) developed a model for the management of guidance in which she identified and described three major management functions, including planning, implementing, and evaluating. Jones, Dayton, and Gelatt (1977) described a program planning model developed at the American Institutes for Research. They outlined the steps involved from the initial planning activities to evaluation report writing. Similarly, Ewens, Dobson, and Seals (1976) outlined the steps involved in developing and implementing a systematic guidance program. Finally, two planning models complete with procedural guides, audio-visual materials, survey instruments, and staff training manuals have been developed at the National Center for Research in Vocational Education in Columbus, Ohio. The *Career Planning Support System* was developed under the leadership of Robert Campbell (1980) and the *Cooperative Rural Career Guidance System* was developed under the leadership of Harry Drier (1979).

A major feature of all these emerging comprehensive guidance programs is *person-centered outcomes*, the achievement of which is the responsibility of the program. Knowledge and skills to be learned by individuals as a result of the program are often grouped in such domains of human growth and development as Self-knowledge and Interpersonal skills; Life Career Planning; and Life Roles, Settings, and Events. In this latter domain subcategories frequently include work and economic understanding as well as leisure and avocational pursuits. These categories serve to identify those domains of human growth and development from which goals and objectives (outcomes) are drawn. The sequencing of these goals and objectives is accomplished according to one of the variety of ways of outlining life stages. Guidance activities and resources designed to assist individuals reach these outcomes are organized accordingly.

Other features of these emerging programs include a *definition* of guidance in outcome terms, *rationale statements* to support the definition, and *assumptions* about human growth and development and about guidance that give shape and direction to the program. Frequently, definitions are stated in outcome terms providing readers with an overview of what knowledge, skills, and understandings individuals can gain from participating in guidance programs as well as indicating how guidance programs relate to other programs in the school. Rationale statements often report the results of needs assessment surveys and statements of authorities concerning the need for guidance and counseling. Closely related to rationale statements are statements of assumptions about human growth and development or the ages, stages, or settings in which the program resides. Such statements help give shape and direction to the program by outlining areas of program responsibility.

In addition, most approaches to guidance programming describe the ways in which guidance activities and resources are organized to reach agreed-upon outcomes. Most, if not all, approaches include the regular *curriculum* as one delivery system. This is done through infusion of guidance goals, objectives, and activities into such disciplines as English, social studies, and science or through the use of mini courses, special classes, or learning packages. In addition, most, if not all, approaches provide for *placement, follow-up, and follow-through* activities to assist individuals in their next step educationally and occupationally. Also, *direct delivery* of guidance activities on a demand basis is a substantial part of most guidance programs. This element is included because there is usually a need for direct immediate services to individuals sometime while they are in contact with the program.

UNDERSTAND NECESSARY CONDITIONS

As changes are being contemplated concerning how guidance activities are organized and implemented, it is important to understand that there are a number of conditions that are prerequisite for successful change to take place. The guidance staff of the Los Alamos, New Mexico, Public Schools found that these conditions included (1) counselors committed to program improvement; (2) counselors committed to change, if change is needed; (3) counselors committed to formulating specific goals; (4) support for the coordinator from administration and counselors; (5) funding for inservice training, and (6) backing from the local school board in the initial stages as well as throughout the program (Engel, Castille, and Neely 1978). This listing is not exhaustive, but it does point to a number of important conditions to be considered as possible changes are being contemplated.

Mitchell and Gysbers (1978) provided a similar list of conditions to be considered for successful transition:

1. All staff members are involved.
2. All staff members are committed to the common objective: total integrated development of individual students.
3. The administration is committed to the comprehensive approach and is willing to negotiate (trade off), helping staff members identify current activities that do not contribute to priority outcomes and supporting staff members' abandonment of such activities in favor of those that do contribute to priority outcomes.
4. All staff members see the comprehensive systematic counseling and guidance program as a function of the total staff rather than the exclusive responsibility of the counselors.
5. Counselors are willing to give up such "security blankets" as writing lengthy reports of their contracts with counselees or seeing counselees individually on matters better addressed in a group.
6. Counselors are interested in acquiring competencies.
7. Staff development activities to help staff members acquire competencies needed for successful implementation of a comprehensive program are provided.
8. Time is made available for planning and designing the program and the evaluation, with all interested groups participating (students, parents, teachers, counselors, administrators, and community).
9. Program developers design an incremental transition rather than abrupt transition that ignores the need for continuing many current activities and thrusts.[2]

Making the transition from guidance as ancillary service to guidance as a comprehensive, developmental program is not easy, automatic, or rapid. It involves changing the behavior patterns of students, parents, the teaching staff, the community, and the guidance staff. Because of this Mitchell and Gysbers (1978) pointed out that while all nine points are important, none is more important than the last. "Abrupt change is difficult and anxiety-producing; it tends to cause participants in the change to build barriers against it."[3]

[2] ©1978 American Personnel and Guidance Association, p. 36. Reprinted by permission.

[3] Ibid.

EXPECT RESISTANCE TO CHANGE

A new conceptualization of guidance is not always accepted enthusiastically by those involved, as Mitchell and Gysbers pointed out. Developmental guidance proponents suggest that the guidance staff should be functioning differently from what they are currently doing, whereas the staff may feel they have already made commitments and their present investment of time and resources can be justified. Their feelings are often expressed in such statements as

---

Comprehensive, developmental guidance is a passing fad!

Look at what success we have had and are continuing to have with our present program!

We should wait until it *really* is better developed!

We are busy 100 percent of our time now!

We could do it if our counselor-student ratio were lower!

It cannot be added on to what is already being done!

Behind these statements lay fears often felt by those faced with change. This human condition of members of the guidance staff must be understood. The failure of a staff readily to embrace a programmatic approach can be appreciated if their original justification for existence and their current functions and operational patterns are understood. Many counselors maintain they are trapped and can react only minimally to change; they are victims of school rigidity and bureaucracy, which places them in quasi-administrative and services functions that impede them from achieving guidance objectives (Aubrey 1973). Elementary school counselors frequently cite their quasi-psychologist functions derived from the educationally disadvantaged emphasis and personal adjustment orientation of the 1960s. This was exacerbated in the 1970s, particularly for elementary school counselors, by the passage of P.L. 94-142, the Education for All Handicapped Children Act, in 1977.

APPRECIATE THE CHALLENGES INVOLVED

The failure of guidance staffs to readily embrace a programmatic approach to guidance can also be better understood if the challenges that the programmatic approach presents are known. Here are some of these challenges:

Guidance programs are based on student outcomes.

Guidance programs are accountable for these outcomes.

Guidance programs are developmental.

Guidance programs are responsive to changing student-school-community needs.

Counselors need to develop new competencies.

Counselors need to consider differential staffing.

Counselors need to be involved in the community.

Counselors need to be involved in the curriculum.

When faced with these challenges, some counselors fear the potential loss of status and power that may come from being associated with the authority of the principal or the school psychologist. Involvement within the community and new relationships with teachers and students may make some counselors uncomfortable. New demands and the need for new competencies threaten others. The most difficult challenge facing most counselors, however, may be the prospect of accepting responsibility for achieving specific student outcomes. Can counselors and guidance programs deliver what they propose?

On the other hand, comprehensive, systematic approaches to guidance have a number of advantages. Campbell (1975, p. 196) suggested five advantages. The first advantage he suggested was that systematic approaches increase the probability that goals will be achieved. Next, he pointed out that a systems approach enables those involved to see the total program. The third advantage was that systems approaches provide for better management and monitoring of the program. Fourth, he pointed out that such approaches help identify alternative ways to achieve goals. Finally, he stated that systems approaches, by definition, have evaluation procedures built into the process.

TRUST DEVELOPMENT NEEDED

Because of the tendency for people to resist change and because of the challenges and risks involved in a programmatic approach, time and privacy are required for a guidance staff as they deal with the issues involved. Time and privacy are required to work through expected resistance to change and to develop trust and working relationships as a total staff. This is true particularly for school districts with guidance personnel at elementary, junior high or middle school, and secondary levels. Current duties at these levels often do not emphasize interaction as much as does this programmatic approach. Thus, the recognition and understanding of the need for counselor trust development is an important first step in the planning process. Provisions for confrontation and the processing of attitudes and feelings in the planning process is mandatory.

FORM WORK GROUPS

Once consensus to change has been reached, the next step is to form work groups to accomplish the tasks involved. Only two standing committees are recommended throughout the entire process: a steering committee and a school-community advisory committee. The majority of tasks can be handled by forming *ad hoc* work groups. The kind of work groups formed will vary depending upon how much the guidance program is being changed. Also, assignments to work groups will vary depending upon the tasks involved. Work groups form and disband as needed.

Steering Committee

The steering committee is responsible for the management of the efforts needed to improve current guidance activities or plan and implement an improved program. This committee is a decision-making body and is responsible for outlining the tasks involved and in making certain that the resources needed to carry out these tasks are available. It monitors the activities of the work group and coordinates, where necessary, the tasks performed by the different work groups.

To carry out these responsibilities, one of the first tasks for the steering committee is to prepare a timetable of the steps it has chosen to take. We recommend the following format: Across the top of the chart, time is represented by months, weeks, or days. Down the side are listed the steps and activities to complete these steps. Since the chart is the master timetable for program change, it requires careful thought and attention. Allow sufficient planning time for this phase of getting organized. Also, keep in mind that the timetable will probably be modified as the program improvement process unfolds.

There are a number of reasons for developing a master timetable. First, it provides those involved with an overview of the scope and sequence of the improvement process. It also shows the relationship of the steps and activities. In addition, potential problems can be identified and therefore anticipated and dealt with in advance. Finally, it provides those involved with an indication of the resources and materials required.

The steering committee should be large enough to reflect a cross-section of the interests of the guidance staff but not so large as to be unwieldy and thus hamper the committee's efficiency. Ordinarily, the steering committee is composed of guidance personnel from each grade level involved. Sometimes an administrator, teacher, parent, or student may also serve. The chairperson of the steering committee is the director or coordinator of guidance. If no such title exists, then the person who ordinarily fills that role should be chairperson.

School-Community Advisory Committee

The school-community advisory committee is composed of representatives from the school and community. The membership on this committee will vary according to the size of the school district and the community but could include such individuals as an administrator (superintendent, assistant superintendent, principal); the guidance director or coordinator; the director of vocational education; a representative of the teaching staff; a representative of the student body; representatives from business, industry, and labor; a representative from the PTA; and a newspaper editor or other media representative.

The school-community advisory committee acts as a liaison between the school and community and provides recommendations concerning the needs of students and the community. A primary duty of the committee is to advise those involved in the guidance program improvement effort. The committee is not a policy or a decision-making body; rather it is a source of advice, counsel, and support. It is a communication link between those involved in the guidance program improvement effort and the school and community. The committee meets throughout the transition period and continues as a permanent part of the improved guidance program. A person outside the school should be chairperson.

The use and involvement of an advisory committee will vary according to the program and the community. It is important, however, that membership be more than in name only, not one that is named at the beginning and then forgotten. An example of the activities of an active advisory committee is the following list of an overall program goal and program objectives for a guidance advisory committee for the Orange County Schools in California.

Advisory Committee for the Improvement of
Guidance Services in Orange County Schools:
Elementary

Overall Program Goal

To raise the percentage of elementary school children in Orange County schools in their progress toward attaining the goals adopted by the Advisory Committee.

Program Objectives

1. To enlist the support and expand the understanding of the decision makers in the educational system as related to the services of the Pupil Personnel Team.
2. To increase the number of districts in Orange County who COMMIT to assisting their children attain these goals.
3. To assist districts develop the processes for raising the percentages of their children registering healthy progress toward these goals:
   a. By designing their guidance programs to make maximum use of existing resources, to establish a plan for adding missing elements, and to continually evaluate and amend the design.
   b. By clarifying the appropriate function of all the people assisting children to reach these goals (e.g., parents, teachers, agency personnel).
   c. By encouraging the development of model programs and the sharing of the results with all Orange County elementary schools.
   d. By encouraging elementary level curriculum changes to include "affective domain" teachings.
   e. By enhancing Orange County elementary school teachers' climate-setting skills.
   f. By assisting parents to enjoy their children.
   g. By encouraging districts to employ Pupil Personnel Services workers in our elementary schools.
   h. By assisting students to develop their life-enhancing skills.
4. To increase the effectiveness of the Pupil Personnel Services workers employed in Orange County elementary schools.
   a. To increase the competencies of the "incumbent" Pupil Personnel Services Workers employed for assisting children reach these goals.

b. To encourage Pupil Personnel Services credentialing requirements to include an assessment as to the candidate's personal suitability for providing the services required.

c. To increase the number of Pupil Personnel Services Workers who are functioning appropriately in their roles.[4]

Most textbooks written on guidance programming have sections dealing with community involvement and community advisory committees. Ours is no exception. Community involvement and interaction are important; there is no doubt about that. However, it is more difficult to accomplish it than it may appear. This conclusion was reached in a study of the impact of the *Rural American Series'* plan to develop and implement career guidance programs.

> The third major conclusion is that involving the community and developing good inter-institutional cooperation is more difficult than it would seem on the surface. In many instances in this field test, limited community involvement and institutional cooperation occurred. Some sites were better at developing linkages between groups and individuals due to the coordinator's skill and possibly due to the existence of prior such linkages. Distance was certainly a factor in inter-institutional cooperation as was perception of institutional roles in regard to the planning of career guidance programs. In many cases, increased positive cooperation which was desirable just did not take place.[5]

This statement was not included to discourage the use of advisory and community involvement committees. Rather it was included to point out that the process is much more complex than many people imagine. Careful planning and continuous effort are required by all involved. It is difficult to do, but the time and effort spent will be well worth it.

[4] Patricia Hooper, "Guidance for an Advisory Committee," Orange Country (California) Department of Education, 1977.

[5] J. W. Altshuld, K. S. Kimmel, V. Axelrod, W. M. Stein, and H. N. Drier, *From Idea to Action: Career Guidance Plans of Rural and Small Schools*, (Columbus, Ohio: The National Center for Research in Vocational Education, The Ohio State University, 1978).

BEGIN PUBLIC RELATIONS ACTIVITIES

Effective public relations activities don't just happen. Careful planning is required. Nor can they be separated from the basic comprehensive, developmental program. In fact, the best public relations begins with a sound, comprehensive, developmental program. The best public relations in the world cannot cover up an ineffective guidance program that does not meet the needs of its consumers.

Public relations planning is a part of the overall program of work in the guidance program improvement process. To be systematic, public relations activities are installed as an ongoing part of the program's improvement and management procedures. Public relations activities that are not related in this fashion to the total program will be seen as superficial and as a result will not have sufficient impact.

To develop your plan for public relations, consider these steps:

---

1. Conduct a thorough appraisal of the public relations resources available in your community.
2. Consider the relative impact each resource may have on various publics.
3. Translate these resources into public relations strategies to be used.
4. Outline the steps that will be taken in the development of these strategies and relate them to the overall plan for program improvement.

Well-planned public relations activities are an integral part of the guidance program improvement process. Remember an effective public relations plan is sincere in purpose and execution; in keeping with the total guidance program's purpose and characteristics; positive in approach and appeal; continuous in application; comprehensive in scope; clear, with simple messages; and beneficial to both the sender and the receiver.

REFERENCES

Altschuld, J. W., K. S. Kimmel, V. Axelrod, W. M. Stein, and H. N. Drier, *From Idea to Action: Career Guidance Plans of Rural and Small Schools.* Columbus, Ohio: National Center for Research in Vocational Education, 1978.

Aubrey, R. F., "Organizational Victimization of School Counselors," *School Counselor*, 20 (1973), 346-54.

Ballast, D. L., and R. L. Shoemaker, *Guidance Program Development.* Springfield, Ill.: Charles C Thomas, Publisher, 1978.

Campbell, R. E., "The Application of Systems Methodology to Career Development Programs," in *A Systems Approach to Learning Environments*, ed. S. D. Zalatino and P. J. Sleeman. Roselle, N.J.: Needed Projects, Inc., 1975.

———. *The Career Planning Support System.* Columbus, Ohio: National Center for Research in Vocational Education, 1980.

Drier, H. N., *Cooperative Rural Guidance System.* Columbus, Ohio: National Center for Research in Vocational Education, 1979.

Engel, E., R. Castille, and J. Neely, "Why Have a Traumatic Time Creating an Accountable Developmental Guidance Program When Somebody Else Already Had That Particular Nervous Breakdown?" Report on an ESEA Title II Project, APGA Convention, Washington, D.C., March 23, 1978.

Ewens, W. P., J. S. Dobson, and J. M. Seals, *Career-guidance–A Systems Approach.* Dubuque, Iowa: Kendall/Hunt Publishing Company, 1976.

Jones, G. B., C. A. Dayton, and H. B. Gelatt, *New Methods for Delivering Human Services.* New York: Human Sciences Press, 1977.

Marland, S. P., Jr. "Counselors an Indispensable Key," *College Board News,* September 1977.

Mitchell, A. M., and N. C. Gysbers, "Comprehensive School Guidance and Counseling Programs," in *The Status of Guidance and Counseling in the Nation's Schools: A Series of Professional Issue Papers.* Washington, D.C.: American Personnel and Guidance Association, 1978.

Orange County Board of Education, *Guidance for an Advisory Committee.* Orange County, Calif.: Orange County Department of Education, 1977.

Ryan, T. A., *Guidance services.* Danville Ill.: The Interstate Printers & Publishers, Inc., 1978.

Chapter 2

Assessing Your Current Program

Current program assessment is a process for obtaining an accurate description of the guidance program as it exists in the present. No attempt is made to judge the value of specific activities. Nor is the program assessment process a way of measuring students needs. Rather, it is a way of determining what the program is like now. The assessment process reveals what the program is accomplishing in its present form. The results are the basis for helping the guidance staff define their program, for educating others about the program, and for planning needed program changes.

There are five steps involved in assessing a current program. As you will see, several steps can be carried out at the same time. Step 1 involves identifying and writing down by grade level all the guidance activities presently being conducted by the guidance and teaching staff. In step 2 the intended outcomes for students are determined for each activity identified in step 1. What competencies will students have as a result of their involvement in an activity? The next step, step 3, requires the identification of the current resources available in the school and community. In step 4 perceptions of the program and the status of students are gathered from students, teachers, administrators, counselors, parents, and community members plus other sources as needed. The fifth step involves the guidance staff keeping a record of their activities and the time spent doing these activities.

To accomplish the current assessment, several work groups will need to be formed. We suggest that steps 1 and 2 be combined and, depending upon the size of school district, that several work groups take on these combined tasks. Concurrently, several other work groups can begin the process of identifying and organizing the current resources available to the guidance staff. At the same time another work group can begin to collect perceptions about the current program. Since all staff are involved in step 5, no work group is needed to collect the information; however, a work group may be needed to tabulate and report the data.

IDENTIFY CURRENT ACTIVITIES AND OUTCOMES

A major task in assessing the current program is to identify and write down by grade level guidance activities in which the guidance and teaching staff are involved. We suggest that one or more work groups devise a form and distribute it to the staff asking them to list guidance activities. The form could contain a few examples by grade level in order to assist staff in knowing what to list. Then the contents from each form is transferred to large sheets of paper appropriately titled grade K, grade 1, and so on through grade 12. Once this is done, the sheets can then be arranged in grade sequence on tables or on a wall so that all of the guidance activities K–12 can be seen together.

There are several reasons for writing these activities on paper. First, the discipline of writing as opposed to speaking is important. Writing forces preciseness. In addition, when these activities are written they are visible to others. This is important because so much of what we do as counselors is invisible to others.

The next task, using the same work groups or, for variety, different work groups, is to identify the student outcomes for each of the activities. Take one activity at a time and ask such questions as, Why do we do this? How are students different as a result of these activities? What do students know or what can students do that they could not do before? Note the following examples:

Activity	*Outcome*
Assist students in planning their schedules.	Students can select classes consistent with their abilities and interests.
Conduct Career Day.	Students can identify an occupation consistent with their abilities and interests.

If the guidance staff believes that these are desirable outcomes, the next question is, How do we know those outcomes were reached by all students for whom the activity was conducted? Although the staff may feel that the activities accomplished the desired outcomes, there may be little evidence to prove it. Thus, the next task is to identify which students attain which outcomes. Is it only the students who drop in to visit the guidance office? Is it all students? Is it all tenth-grade students? Indicate the number and percent of the students achieving this outcome and how it was determined that this outcome has been achieved. In this way the guidance staff is defining the outcomes for specific subpopulations in the school. Follow this technique through with each activity (or grouping of activities),

Activity	*Outcome*	*Impact*
Assist students in planning their schedules.	Students can select classes consistent with their abilities and interests.	1. Only 12 percent of the students change classes. 2. 1976 follow-up study of graduates: 68 percent reported curriculum satisfaction.
Conduct Career Day.	Students can identify an occupation consistent with their abilities and interests.	1. Participation-attendance records at the Career Day: low (only 30 percent). 2. Counselor contacts: high-quality response but only 12 percent of the student body.

and the result will be a listing of the outcomes that the activity is achieving and the impact it is having on students. It may be that many guidance outcomes are attained by only a small number of students, or there may be little proof that any outcomes are attained at all.

Data from a current program assessment provide information that reveals what the guidance program is accomplishing in its present state. Data from the current program assessment can help indicate areas of success or suggest where revisions are necessary or where new activities may be needed. The assessment also provides information about how guidance personnel spend their time—the tasks completed, the outcomes achieved.

Then you can suggest or actually make changes in schedules to conform to the desired outcomes. You can establish priorities and plan activities in accordance with them. For example, the assessment may indicate that schedules are dominated by student conferences but that the work actually affects only a small percentage of students. It may be decided that there is a need to review the total operation, to do more work with groups, or to devote more time to teachers or parents. Guidance program decisions will be influenced by these comparisons.

IDENTIFY CURRENT RESOURCES

The first task is to determine the nature of the resources available. DuBois (1975) suggested there are human, financial, and technical resources and that these terms can be used to categorize available resources.

Human Resources

School staff members are important members of the guidance team, and as such their competencies and their potential contributions need to be better understood. Staff competencies and their potential contributions to the guidance program need to be identified, and any direct participation should be logged. Students are also important members of the guidance team. Working as aides in career centers, preparing resource units, or making presentations about jobs they have held are just a few of the ways students contribute. Such contributions need to be identified and logged in a similar way to staff participation.

One of education's chronically untapped resources is the community; therefore, an inventory of the community is a major component of any resource assessment. While it is not done every year, when it is done, it should be carried out systematically and comprehensively.

One way to begin an inventory of the community is to make initial contacts by phone. Then follow-up letters are sent. The letter should describe the guidance program in the district, explain what the role of community members could be, and ask for their participation in the program. Forms should be enclosed for them to fill out and return.

Financial Resources

The place to begin the current assessment of financial resources is the current budget of the guidance program. Even if there is no official budget, some funds are being spent for guidance activities, so begin there. List such items as salaries for counselors, guidance secretaries, or aides; supplies such as paper, pencils, record folders; guidance materials such as books, filmstrips, films, records, pamphlets; standardized tests; and services such as test scoring. All expenses for guidance activities in the district should be included. Also check to see if any money from federal, state, or private foundations is being used.

Technical Resources

While you complete the technical resources assessment part of the overall assessment, ask such questions as, What standardized tests are provided? How many copies? What questionnaires or inventories are used? What infomation is available from student records? What guidance materials—for example, guidance references, books on careers and colleges, filmstrips, records, films, pamphlets—are available? Data about the guidance material to be gathered should include: amount of each piece of equipment or material, title, copyright date, brief description of content, the grade level(s) for which they can be used, and the career education component(s) they address. The inventory can also include information about how to obtain the resources and about any restrictions on borrowing.

Resources Catalog

After information about available resources has been gathered, the next step is to organize the information and to catalog it in a way that will be most useful to members of the guidance staff. It is suggested that a loose-leaf notebook be used to organize the forms that are completed. This information also can be kept on three-by-five cards. When additional resources are acquired, the information can easily be added to the loose-leaf binder and the card file.

GATHER PERCEPTIONS

A major step in the current assessment process, step 4, is the gathering of perceptions about the current program from students, teachers, counselors, administrators, parents, and community members. In addition, data about the current status of students is also collected and analyzed at this time. This phase of the current assessment focuses on what individuals from such groups as these think about the activities of the current program. It does not focus on the perceived needs of individuals, the school, or society. That kind of assessment comes later.

Gathering Perceptions about the Program

The most direct way to find out what people think about your current program is to ask them. One way to do that is to interview a representative sample of individuals from the various groups involved using a structured interview approach. One or more staff work groups should be formed to accomplish this task. The advantages of using an interview approach are the direct contacts with members of the various groups and the in-depth responses that can be gathered. The disadvantages include the time-consuming nature of the task, the small numbers of people that can be contacted, and the difficulty in tabulating the results.

Another way to gather perceptions is to use a questionnaire. Questionnaires can be prepared and distributed to large numbers of people. Tabulation of results is easy. On the other hand, the personal touch of the interview is lost, as is the opportunity to gain in-depth responses. Because of the trade-offs involved you may wish to consider some combination of interviews and questionnaires to accomplish the task.

The following is a sample of a partial questionnaire to be used with students. It was adopted from a questionnaire in a training module developed in the Mesa Public Schools, Mesa, Arizona (Richins and Scott 1974).[1]

Guidance Program Questionnaire

Directions: Please read each statement carefully. There are no right or wrong answers. Just check the space that best decribes how you feel about each statement concerning the guidance activities in your school.

The guidance program helped me to:	*Strongly Agree*	*Agree*	*No Opinion*	*Disagree*	*Strongly Disagree*
1. Plan my schedule	______	______	______	______	______
2. Interpret tests and other data	______	______	______	______	______
3. Recognize my abilities	______	______	______	______	______
4. Solve personal problems	______	______	______	______	______
5. Plan for education after high school	______	______	______	______	______

[1] Courtesy of Guidance Department, Mesa Public Schools, Mesa, Arizona. Byron E. McKinnon, Director.

6. Make career plans	______	______	______	______	______
7. Seek employment	______	______	______	______	______
8. Seek financial aid	______	______	______	______	______

The format of this questionnaire can be used with all of the groups surveyed by changing the statement "The guidance program helped me to . . ." For example, if the questionnaire were to be used with parents, the statement could read, "The guidance program helped my son or daughter to . . . " In addition, the activities list can be changed to reflect the activities of your own program. It is recommended that a questionnaire of this type be prepared for the elementary, junior high or middle school, and senior high levels. The range of response also can be modified to three—agree, no opinion, disagree—or simply yes or no.

Gathering Student Status Information

An important part of current program assessment is gathering information about the current status of students. A work group should take on the task of identifying and collecting data about students. Possible sources of such data include standardized tests, criterion-referenced tests, attitude surveys, follow-up studies, and dropout studies. When compiled, such information can provide insight into how well the current program is meeting students needs.

Until a few years ago most of these measures were indirect measures of the outcomes of the guidance program because they were not tied directly to the goals and objectives of the program. Most were instruments that measured ability, aptitude, interests, and personality, or they were achievement measures in such areas as English, history, and math. That is less true today because of the recent development of instruments that measure career maturity and achievement in such guidance content areas as decision making.

In addition to providing insight into how well the current program is meeting student needs, student data play another role in the program development. They provide a baseline against which to compare students in future years. They provide the opportunity, in later years, to look at trends concerning student growth and development.

KEEP TRACK OF STAFF TIME AND ACTIVITIES

An important part of the current assessment is the time-and-activity log kept by members of the guidance staff. The log should cover a long enough time period so that some activities that may occur only occasionally are not given undue attention. DuBois (1975) recommended that the school year be divided into quarters and that nine days be chosen at random for each quarter. If information is needed quickly, then the best time for you should be chosen. Try for at least one week and remember to correct for the possible distortion that could occur in a one-week sample.

DuBois (1975) recommended that a simple format be used.

> The format of the log should not be so complicated that it is a burden to use. However, the form should include this information:
>
> What happened?
> Who was involved?
> How long it took.
> Why it was done. (p. 27)

The following is an example of how this format might be used:

Date	*Time*	*Activity*	*Purpose*
9/29/79	1:00-1:30	visit with parent	discuss work plans of son
	1:30-1:45	visit with new student	get acquainted session
	1:45-2:30	met with 9th grade Social Studies class	decision-making exercises

Once such data are collected for the appropriate length of time, the next step is to categorize the activities and the purposes so that the general thrust of the current program can be understood. This can be done using preshaped charts, bar graphs, or simple percentages. For example, the Mesa, Arizona, Public Schools[2] reported the time spent at the elementary, junior high, and senior high levels as follows:

Current Elementary School Program Summary

I.	Academic learning counseling	10%
II.	Educational-vocational counseling	5%
III.	Interpersonal counseling	50%
IV.	Intrapersonal counseling	20%
V.	Conducting student registration	2%
VI.	Handling attendance concerns	3%
VII.	Conducting and receiving in-service training	10%
VIII.	Supervising clubs	0%

Current Junior High School Program Summary

I.	Academic learning and counseling	12%
II.	Educational-vocational counseling	10%
III.	Interpersonal counseling	36%
IV.	Intrapersonal counseling	27%
V.	Providing school support—such as conducting student registration and handling noncounseling activities	10%
VI.	Providing teacher-staff counseling	3%
VII.	Conducting community activities	2%

Current High School Program Summary

I.	Academic learning counseling	5%
II.	Educational-vocational counseling	10%
III.	Interpersonal counseling	5%
IV.	Intrapersonal counseling	5%

[2] Courtesy of Guidance Department, Mesa Public Schools, Mesa, Arizona. Byron E. McKinnon, Director.

V.	Conducting student registration	66%
VI.	Handling attendance concerns	5%
VII.	Conducting and receiving in-service training	3%
VIII.	Supervising clubs	1%

REFERENCES

DuBois, P., *Module 4: Assessing Current Status.* Palo Alto, Calif.: American Institutes for Research, 1975.

Mesa Guidance and Counseling Department, *Toward Accountability.* Mesa, Ariz.: Mesa Public Schools, 1973.

Richins, D. and S. Scott, *Current Guidance Program Assessment.* Mesa, Ariz.: Mesa Public Schools, 1974.

Chapter 3

Selecting and Using a Developmental Guidance Program Model

Selecting and using a model is a major task in your program improvement process. The model serves as an ideal against which you can compare your current program; it serves as a template to lay over the top of the current program so that similarities and differences can be seen. Having made this comparison, you can then *adopt* those components of the model that fit your situation, *adapt* other components of the model as needed, or *create* new components because of unique local needs. For our purposes the model we will use to illustrate the major components of a comprehensive developmental guidance program is the Career Conscious Individual Model.

THE CAREER CONSCIOUS INDIVIDUAL MODEL

The Career Conscious Individual Model for guidance and education was conceptualized by Gysbers and Moore (1971). It is an outcome-oriented model, designed to provide a comprehensive overview of the knowledge, skills, and attitudes (competencies) individuals need in order to facilitate their development. The concept of consciousness is taken from the work of Reich (1971) and his description of how consciousness functions in individuals.

> Included within the idea of consciousness is a person's background, education, politics, insight, values, emotions, and philosophy, but consciousness is more than these or even the sum of them. It is the whole man; his "head"; his way of life. It is that by which he creates his own life and thus creates the society in which he lives.[3]

In helping individuals reach their potential, educational programs are stimulating career consciousness—the ability for individuals to visualize and plan their lives. The challenge to education and guidance is to create career consciousness in individuals: to help them project themselves into possible life roles, settings, and events; analyze these roles, settings, and events; and relate their findings to their present situations.

Life Career Development

The Career Conscious Individual Model is based upon life career development concepts and principles. Life career development is defined as self-development over the life span through the integration of the roles, settings, and events in a person's life. The word *life* in the definition indicates that the focus of this conception of human growth and development is on the total person—the human career. The word *career* identifies and relates the many and often-varied roles in which individuals are involved (student, worker, consumer, citizen, parent); the settings in which individuals find themselves (home, school, community); and the events that occur over their lifetimes (entry job, marriage, divorce, retirement). Finally, the word *development* is used to indicate that individuals are always in process of becoming. When used in sequence, the words *life career development* bring these separate meanings together, but at the same time a greater meaning evolves. Life career development describes total individuals: unique individuals, each with their own life-style.

[3] From *The Greening of America*, by Charles A. Reich. Copyright © 1970 by Charles A. Reich. Reprinted by permission of Random House, Inc. A portion of this book originally appeared in *The New Yorker* in somewhat different form.

In the definition of life career development, the word *career* has a substantially different meaning from some definitions. Here it focuses on all aspects of life, not as separate entities but as interrelated parts of the whole person.

> The concept of *career* encompasses a variety of possible patterns of personal choice related to each individual's total life style . . .
>
> 1. occupations
> 2. education
> 3. personal and social behavior
> 4. learning-how-to-learn
> 5. social responsibility (i.e., citizenship)
> 6. leisure time activities
>
> (Jones, Hamilton, Ganschow, Helliwell, and Wolff 1972, p. 6)

Super (1976) suggested an equally broad definition of the career concept. He defined career as

> the sequence of major positions occupied by a person throughout his preoccupational, occupational, and postoccupational life: includes work-related roles such as those of student, employee, and pensioner, together with complimentary avocational, familial, and civic roles. Careers exist only as people pursue them; they are person-centered. (p. 20)

It should be clear that the term *career*, when viewed from this broad perspective, is not a new word for occupation. People have careers, the work world or market place has occupations. Unfortunately, too many people use the word *career* when they should use the word *occupation*. Also the term *career* is not restricted to some people. All people have a career; their life is their career. Finally, the words *life career development* do not delineate and describe only one part of human growth and development. While it is useful to focus at times on different aspects of development–physical, emotional, and intellectual, for example–there is also a need to integrate these aspects of development. Life career development is advocated as an organizing and integrating concept for understanding and facilitating human growth and development.

Career Conscious Individual Model Elements

The Career Conscious Individual Model focuses on self-development as effected by four interrelated domains or dimensions of human growth and development: *Self-knowledge and Interpersonal Skills; Life Roles, Settings, and Events; Life Career Planning*; and, *Basic Studies and Occupational Preparation*. These four domains are discussed briefly in the following paragraphs.

Self-knowledge and Interpersonal Skills

In the Self-knowledge and Interpersonal Skills Domain the focus is on helping individuals understand themselves and others. The main concepts of this domain involve individuals' awareness and acceptance of themselves, their awareness and acceptance of others, and their development of interpersonal skills. Within this domain individuals begin to develop an awareness of their personal characteristics–interests, aspirations, aptitudes, and abilities. Individuals learn techniques for self-appraisal and the analysis of their personal characteristics in terms of a real-ideal self-continuum. They begin to formulate plans for self-improvement in such areas as physical and mental health. Individuals become knowledgeable about the interactive relationship of self and environment in such a way that they develop personal standards and a sense of purpose in life. Individuals learn how to create and maintain relationships and develop skills that allow for beneficial interaction within those relationships. Outcomes of this domain are persons who can use self-knowledge in life career planning, who have positive interpersonal relations, and who are self-directed in that they accept responsibility for their own behavior.

Life Roles, Settings, and Events

The emphasis in this domain is on the interrelatedness of various life roles (for example, learner, citizen, consumer), settings (for example, home, school, work, and community), and events (for example, job entry, marriage, retirement) in which individuals participate over the life span. Emphasis is given to the knowledge and understanding of the sociological, psychological, and economic dimensions and structure of their world. As individuals explore the different aspects of their roles, they learn how stereotypes affect them and the lives of others. The implication of futuristic concerns are examined and related to their lives. Individuals learn of the potential impact of change in modern society and of the necessity of being able to project themselves into the future. In this way they begin to predict the future, foresee alternatives they may choose, and plan to meet the requirements of life career alternatives. As a result of learning about the component dimensions of their world, individuals learn of the reciprocal influences of life roles, settings, and events with life-style preferences.

Life Career Planning

The Life Career Planning Domain is designed to help individuals understand that decision making and planning are important tasks in everyday life and to recognize the need for life career planning. Individuals learn that there are many occupations and industries comprising the work world and that they can be grouped according to occupational requirements and characteristics and personal skills, interests, values, and aspirations. Emphasis is placed on individuals' learning of various rights and responsibilities associated with their involvement in a life career.

The central focus of this domain is on the mastery of decision-making skills as a part of life career planning. Individuals develop skills in this area by identifying the elements of the decision-making process. They develop skills in gathering information from relevant sources, both external and internal, and learn to use the collected information in making informed and reasoned decisions. A major aspect of this process involves the appraisal and application of personal values as they may be related to prospective plans and decisions. Individuals engage in planning activities and begin to understand that they can influence their future by applying such skill. They begin to accept responsibility for making their own choices, for managing their own resources, and for directing the future course of their own lives.

Basic Studies and Occupational Preparation

The fourth domain, Basic Studies and Occupational Preparation, forms the basis for the overall instructional program. It is the largest component in terms of amount of content and number of activities. This domain contains the knowledge, skills, and understandings found in such education disciplines as English, social studies, mathematics, fine arts, industrial arts, home economics, physical and health education, foreign language, and vocational-technical education.

These areas of education are basic to the total development of individuals, but because of new and emerging challenges to education they now need to be viewed in new interdisciplinary ways. The roles, settings, and events of a person's life career and the interrelated worlds of education, work, and leisure can serve as a primary content focus for knowledge acquisition and skill development in basic studies and occupational preparation. Also, since the education, work, and leisure worlds are undergoing constant change, individuals need to update their knowledge and skill in basic studies and occupational preparation. Thus, a necessary emphasis within this domain involves the continuous acquisition and refinement of basic and occupational knowledge and skills throughout life.

TWO MAJOR DELIVERY SYSTEMS

Education as envisioned by the Career Conscious Individual Model includes two major, interrelated delivery systems: the instruction program and the guidance program. Each delivery system emphasizes separate specific learning outcomes, but at the same time there is overlap. The instruction program's outcome goals, objectives, and student competencies are grouped under the Basic Studies and Occupational Preparation Domain using such titles as fine arts, vocational-technical education, science, physical education, mathematics, social studies, foreign language, and English. The guidance program's outcome goals, objectives, and student competencies are grouped under the domains Self-knowledge and Interpersonal Skills; Life Roles, Settings, and Events; and Life Career Planning. The instruction program is by far the largest in terms of numbers of goals, objectives, and student competencies, but it is no more important than the guidance program. That is why the circles in Figure 3.1, which depict the

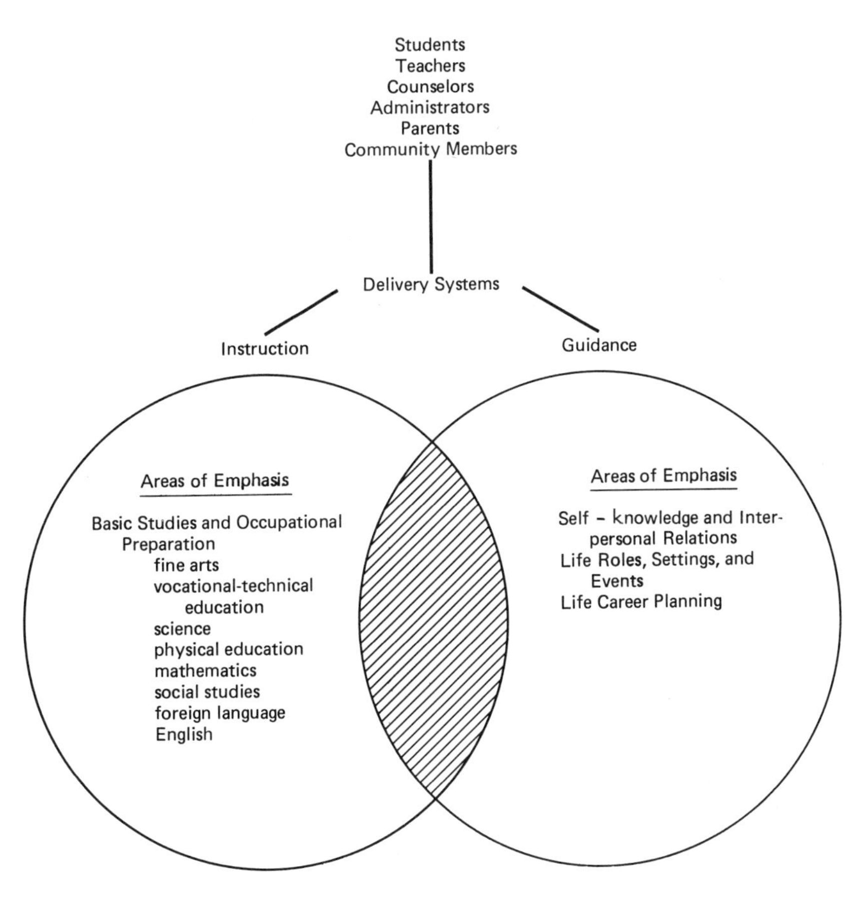

FIGURE 3.1
Career Conscious Individual Model Delivery Systems

delivery systems, are equal in size. Figure 3.1 also illustrates the fact that there are separate learnings in each system (nonshaded area), which require specific attention. At the same time these learnings overlap (shaded area), requiring that the instruction program support guidance program elements at times and at other times requiring that the guidance program support the instructional program. It is not a case of either/or but of both/and.

Richardson and Baron (1975) outlined a similar schema but labeled their systems as two major purposes of education: social learnings and personal learnings. They pointed out that teachers were primarily responsible for the instructional function—that of "guiding the learning of developmental tasks in the area of 'social learnings' "[4] –while counselors were primarily responsible for the counseling function–that of "guiding the learning of developmental tasks in the areas of the 'personal learnings' "[5]. They also pointed out that the counseling program contains developmental and crisis emphases, whereas the instructional program focuses on developmental and remedial elements.[6]

CAREER CONSCIOUS INDIVIDUAL GUIDANCE PROGRAM COMPONENTS

Whatever guidance program model you select, we recommend that it contain the following seven components:

Structural Components
1. Definition
2. Rationale
3. Assumptions

Program Components

4. Guidance Curriculum
5. Individual Planning
6. Responsive Services
7. System Support

The first three components tell readers about the competencies individuals will possess as a result of the program and where the program fits in relation to other educational programs (definition), reasons why the program is important (rationale), and the

[4] H. D. Richardson and M. Baron, *Developmental Counseling in Education,* Guidance Monograph Series © 1975, p. 21. Reprinted by permission of TSC.

[5] Ibid.

[6] Ibid., p. 59

premises upon which the program rests (assumptions). We have labeled these components structural components. The next four components delineate the major activities, roles, and responsibilities of personnel involved in the guidance program. Hence we have labeled them program components. These include the guidance curriculum, individual planning, responsive services, and system support.

Structural Components

Definition

A definition of guidance identifies the centrality of guidance within the educational process and delineates, in broad outcome terms, the competencies individuals will possess as a result of their involvement in the program. An example of a definition of guidance is as follows: Guidance is an integral and central, but specifically identifiable and accountable, part of the total continuing education process. It is a comprehensive and developmental educational program responsible for assisting all individuals in developing positive self-concepts, effective human relationships, decision-making competencies, and understanding of current and potential life roles, settings, and events; it is also responsible for placement competencies to aid them in the transition from one setting to another. It is also responsible for assisting all individuals to understand the meaning of basic studies and occupational preparation and relate them to their present and future lives.

Rationale

A rationale discusses the importance of guidance as an equal partner in the educational process and provides reasons why individuals in our society need to acquire the competencies that will accrue to them as a result of their involvement with a comprehensive developmental guidance program. An example of a rationale for a program follows. Note that it is based on the foregoing definition.

The first part of the foregoing definition emphasizes the *centrality* of guidance in education. At the same time it stresses the fact that guidance is an *identifiable and accountable* program. This means that guidance is a comprehensive developmental program based on individual and societal needs organized around person-centered goals and activities designed to meet those needs. More specifically, this means that guidance is an educational program commensurate with other major educational programs in the school.

The second part of the definition stresses the point that guidance as a *comprehensive, developmental educational program* places emphasis on individual development—in other words, an educational program responsible for assisting all individuals. Individuals today face depersonalization in many facets of their lives as bureaucracies and impersonal relations are commonplace. They often feel powerless in the face of masses of people, mass communication, and mass everything else and need help in dealing with these feelings, not at the expense of society but in the context of society. Their feelings of control over their environment and their own destiny and their relations with others and institutions are of primary importance in guidance programs. Individuals must be viewed as totalities, as individuals. Their development can best be facilitated by guidance programs that begin in the early years and continue to be available throughout the lives of individuals.

The third part of the definition identifies five specific areas of human growth and development for which guidance programs are responsible. The first area in the definition is that of assisting all individuals in developing *positive self-concepts and effective human relationships.* This means that a major emphasis in guidance programs is on individuals learning about themselves, learning about others, and learning about interactions between self and others. Formerly, individuals were brought up in a fairly stable society in which their roles were defined and relationships with others were fairly constant. Now they face an increasingly mobile society, in which relationships with both people and things are becoming less and less enduring. Society is characterized by transience and impermanence. Traditional beliefs and ways of doing things no longer seem sufficient for coping with the environmental demands. As a result, many individuals today have problems defining their roles; thus, the quests for answers to Who am I? and Where do I fit in? Guidance programs can help individuals respond to these questions.

The development in individuals of self-appraisal and self-improvement competencies are a primary goal of guidance. Through learnings in this area individuals become more aware of personal characteristics such as aptitudes, interests, goals, abilities, values, and physical traits and the influence these characteristics may have on the persons they are and can become. Being able to use self-knowledge in life career planning and interpersonal relationships and to assume responsibility for one's own behavior are examples of desired outcomes.

The second area of human growth and development for which guidance is responsible is the development of *decision-making competencies* in individuals. The need for attention to this area is well documented (Prediger, Roth, and Noeth 1973), since planning for and making decisions are vital tasks in the lives of individuals. Everyday decisions are made that influence one's life career. Mastery of decision-making skills and the application of these skills to life career planning are central learnings within this area of guidance. A preliminary task to effective decision-making is the clarification of personal values. The degree of congruence between what individuals value and the outcomes of decisions individuals make, contribute to personal satisfaction.

Individuals learn within this area to identify the steps necessary in making decisions. Included are the skills for gathering and using relevant information. Understanding the influence of planning on one's future and the responsibility one must take for planning are components of the life career planning process. Life career planning is ongoing. Change and time affect one's planning and decisions. A decision outcome that is satisfactory and appropriate for the present may, with time or change, become unsatisfactory or inappropriate. Thus, the ability to evaluate decisions in view of new information or circumstances is vital. Being able to clarify personal values, identify steps needed to make personal decisions, gather relevant information, and apply decision-making skills to life career plans are examples of desired outcomes in this area of guidance.

The third area for which guidance is responsible is assisting all individuals in developing understanding of *current and potential life roles, settings, and events.* Increasing societal complexity affects not only interpersonal relationships and feelings of individuality but also other life roles, settings, and events, specifically including those associated with the worlds of education, work, and leisure. Changes resulting from advances in technology are perhaps more apparent as they affect the world of work. No longer are individuals well acquainted with the occupations of family and community members or their contributive roles to the common good of the society. Parents' occupations are removed from the home and often from the immediate neighborhood.

Because individuals over their lifetimes will be assuming a number of roles, functioning in a variety of settings, and experiencing many events, learnings in this area emphasize their understanding of the various roles, settings, and events that interrelate to form their life careers. The roles of family member, citizen, worker, and leisure participant; settings such as home, school, community, and work; and events such as birthdays, educational milestones, job entry, and job change are identified and examined in terms of their influence on life-styles. Learnings in this area include developing understanding of the structure of the family, education, work, and leisure requirements and characteristics. The effect of change—natural as well as unexpected, social as well as technological, in self as well as in others—is a major learning. Being able to effectively answer personal identity questions contained in the questions Who? (roles), Where? (settings), and When? (events) are desired outcomes in this area of guidance.

Guidance is also responsible for assisting all individuals in developing *placement competencies to aid them in the transition from one setting to another.* As individuals move from one setting to another, they need specific knowledge and skills in order to make such moves as effectively as possible. While placement is defined broadly, specific attention is given to intra- and intereducational and occupational transitions and to the personal competencies needed to make such transitions. Personal competencies include knowledge of the spectrum of educational courses and programs, an understanding of the relationships they may have to personal and societal needs and goals, and skills in using a wide variety of information and resources. It also includes an understanding of the pathways and linkages between those courses and programs and potential personal goals. Stress is placed on employability skill development including résumé writing, job searching, and job interviewing. Closely tied to these competencies and the program elements required to develop and implement them are the follow-up and the follow-through components of placement. Finding out what happened to individuals as they move from one course, program or job-occupation to another and

providing follow-through assistance as needed are important aspects of guidance.

Finally, guidance is responsible for assisting all individuals to *understand the meaning of basic studies and occupational preparation and relate them to their present and future lives.* Some of the dissatisfaction of youth with education stems from the feeling that what they are doing in school is not relevant to their lives. It is a responsibility of the guidance program to seek to create relevance in the schools and to show individuals how the knowledge, understandings, and skills they are obtaining and the courses they are taking will help them as they progress through their life career.

Assumptions

Assumptions identify and briefly describe the premises upon which a guidance program rests. Assumptions provide the program with its shape and direction, its nature and structure. Examples of assumptions are as follows:

Assumption 1: Guidance programs help develop and protect students' individuality. Guidance personnel and programs have a major responsibility in the educational system to develop and protect the individuality of students. This means that assistance must be provided to all individuals so that they can become aware of their needs and can develop and pursue immediate and long-range personal goals.

Assumption 2: Guidance programs are available to all students at all educational levels. Guidance personnel and programs have the responsibility of serving all students at each educational level, rather than only a selected group of individuals at one level. This means that by design and operation, guidance programs are functioning effectively at each educational level and are part of the educational program of all individuals.

Assumption 3: Guidance programs are lifelong, dealing with developmental as well as prescriptive and remedial concerns. Guidance personnel and programs have the responsibility of meeting the developmental guidance needs that are remedial in nature. This means that guidance programs have a developmental focus that maximizes the prevention of problems and a prescriptive emphasis to assist individuals alleviate continuing concerns.

Assumption 4: Guidance programs are integrated with the total educational process. Guidance goals, objectives, and procedures are integral, central, but still identifiable components of the total educational process. This means an equal and complementary relationship exists between the instructional program and the guidance program. It also means that all educational staff have guidance responsibilities. Program coordination is assumed by the guidance staff in addition to their direct service functions to students.

Assumption 5: Guidance programs are evaluated periodically for effectiveness. If guidance programs and personnel are to be responsive to the guidance needs of those who are served, periodic program and personal evaluation is necessary. To accomplish this will require that guidance programs be organized and implemented from an evaluation perspective.[7]

Program Components

An examination of the emerging needs of students, the variety of guidance methods, techniques, and resources available, and the increased expectations of policymakers, funders, and consumers indicates that a new organizer for guidance programs in the schools is needed. The traditional formulations of guidance—the six services (orientation, information, assessment, counseling, placement, and follow-up)—and the three aspects of guidance (educational, personal-social, and vocational), though once sufficient, are no longer adequate ways to organize guidance programs in today's schools.

When described as services, guidance is often cast as ancillary and is seen as only supportive to curriculum and instruction, not as equal and complementary. The three-aspects view of guidance has frequently resulted in fragmented and event-oriented activities and, in some instances, in the development of separate kinds of programs and counselors. Educational guidance is stressed by academic-college personnel, personal-social guidance becomes the territory of mental health workers, and vocational guidance becomes the focus of vocational education and manpower-labor economists.

[7] The assumptions listed in this section were adapted from material developed at Mesa, Arizona, in a paper titled *Toward Accountability* (Mesa Guidance and Counseling Department 1973) and from material from the American Institutes for Research (Sanderson and Helliwell 1975).

If the proposition that these traditional organizers are no longer adequate is acceptable, then the question is, What is an appropriate one? One way to answer this question is to ask what should be expected of a comprehensive guidance program.

---

1. Are there knowledges, skills, and attitudes needed by all individuals that should be the instructional responsibility of guidance programs?
2. Do individuals have the right to have someone in the school system sensitive to their unique life career development needs, including needs for placement and follow-through?
3. Should guidance staff be available and responsible to special or unexpected needs of students, staff, parents, and the community?
4. Does the school program and staff require support that can best be supplied by guidance personnel?

An affirmative response to these four questions implies an organizer that is different from the traditional services or aspects model. In addition, a review of the variety of guidance methods, techniques, and resources available today and an understanding of the nature of the expectations of policymakers and consumers of guidance also suggest a model different from the traditional services or aspects model. The organizer that is suggested from an affirmative answer to the four questions and from a review of the literature is a program model of guidance techniques, methods, and resources containing four interactive components: a guidance curriculum, individual planning, responsive services, and systems support.

The curriculum component was chosen because a curriculum provides a vehicle to impart to all students guidance content in a systematic way. The next component, individual planning, was included as a part of the model because of the increasing need for all individuals continuously to monitor and understand their growth and development; to consider and take action on their next steps educationally or occupationally. The responsive-services component was included because of the need in comprehensive guidance programs to respond to the direct, immediate concerns of individuals whether these concerns involve information, crisis counseling, or consultation with parents, teachers, or other specialists. Finally, the systems-support component was included because it was recognized that, for the other guidance processes to be effective, a variety of support activities, such as staff development, testing and research, and curriculum development, are required.

These components, then, can serve as an organizer for the many guidance methods, techniques, and resources required in a comprehensive guidance program. In addition, however, they also can serve as a vehicle to check the comprehensiveness of a program. In our opinion a program would not be comprehensive unless it had activities in each of the components.

Guidance Curriculum

One of the assumptions upon which our conception of guidance is based is that there is guidance content that all students should learn in a systematic, sequential way. This means counselor involvement in the curriculum; it means a guidance curriculum. This is not a new idea, for as we said previously, the notion of a guidance curriculum has deep, historical roots. More recently Rice (1966), Swan (1966), Hansen (1972), and Eisenberg (1974) stressed the importance of involving guidance and guidance practitioners in the curriculum. What is new, however, making today different from the past, is the array of guidance and counseling techniques, methods, and resources currently available that work best as a part of a curriculum. What is new, too, is the concept that a comprehensive guidance program has an organized and sequential curriculum.

To establish a curriculum for guidance, it is first necessary to identify those areas of human growth and development that could form a base for guidance from which content could be drawn. Career development theory and research is suggestive of such areas, and is the work of psychologists who have delineated developmental tasks. So, too, is the work of individuals who emphasize psychological, moral, process, and values education.

The recent work of a number of career development theorists, researchers, and practitioners, however, is of particular value in the identification of a content base for guidance programs. Super (1975) emphasized this point when he stated that a theoretical base is available for guidance in the form of career development theory—a theory that describes and relates life stages and developmental tasks career patterns, and individual differences. In a discussion of these factors he (Super 1976) described the concept of life roles, life theaters, and life space as a way of organizing human growth and development through life stages over the life span.

Building on the work of Cole (1972, 1973), Goldhammer and Taylor (1972), Goldhammer (1975), Super (1975), and others, Bailey (1976) identified four general life roles he felt gave operational meaning to Cole's (1973) concept of the educated person. These four roles included work, family, learning and self-development, and social citizenship. To the life roles Gysbers and Moore (1974) added life settings and life events, suggesting that human growth and development over the life span can be described in terms of the interaction of these three constructs.

Previously Herr and Cramer (1972) had described the use of career development theory as a base for guidance programs when they outlined objectives for individuals in such areas as self-knowledge, knowledge of the education-work world, and decision making. In a similar way Hansen and Tennyson (1975) and Tennyson, Hansen, Klaurens and Antholz (1975) outlined a career development curriculum that has self-development as its major focus. The curriculum delineated life stages and the corresponding developmental tasks related to those stages and describes activities that facilitate the accomplishment of these tasks. Later Dudley and Tiedeman (1977) traced the evolution of the Harvard Studies in Career Development, outlining a shift in emphasis from prediction to an understanding of self and decision-making processes. They described the Miller-Tiedeman-Tiedeman cubistic model of decision making, which brings "together three basic elements—psychological, problem, and self-conceptualizing states" (p. 296).

An analysis of the work of these and other theorists, researchers, and practitioners in career development as well as the work of those who emphasize psychological, moral, process, and values education reveals three recurrent themes common to all. Most if not all stress the need for attention to self-concept development—knowledge of self as well as the development of interpersonal skills. Another theme that is common is the concern for assisting individuals in understanding the decision-making process and in developing decision-making skills. Finally, a third common theme is the need for individuals to gain an understanding of current and potential life roles, settings, and events. Richardson and Baron (1975) grouped these three common themes under the term *personal learnings*—learnings that stress developing a self-identity, answering questions dealing with self in terms of who, where, and when. They pointed out that the facilitation of these learnings in individuals is the second major purpose of education (the other being social learnings) and that these personal learnings form the basis for a developmental counseling program.

The guidance domains of the Career Conscious Individual Model provide the body of knowledge and skills from which the guidance curriculum presented in this book was derived. The major content areas of the guidance curriculum are Self-knowledge and Interpersonal Skills, Life Roles, Settings, and Events, and Life Career Planning. These three domains are closely related to the three common themes that emerged from a review of the work of career development theorists, researchers, and practitioners and from individuals who advocate psychological, moral, process, and values education.

Before an outline and selected examples of the curriculum are presented, however, an explanation of how the curriculum was constructed is necessary, since it is somewhat different from other published guidance curricula. It was based in part on the taxonomy of Wellman and Moore (1975), which in turn integrated features of Bloom's (1956) taxonomy and Krathwohl, Bloom, and Masia's (1964) taxonomy.

The Wellman and Moore taxonomy begins with a number of assumptions about human growth and development. Several of these assumptions are as follows:

---

1. Individual development is a process of continuous and sequential, but not necessarily uninterrupted or uniform, progress toward increased effectiveness in the management and mastery of the environment for the satisfaction of psychological and social needs.
2. The stage, or level, of individuals' development at any given point is related to the nature and accuracy of their perceptions, the level of complexity of their conceptualizations, and the subsequent development rate and direction. No individual in an educational setting is at a zero point in development; hence change must be measured from some relative point rather than from an absolute.
3. Positive developmental changes are potential steps toward the achievement of higher-level purposive goals. This interlocking rela-

tionship dictates that achievement at a particular growth stage be viewed as a means to further development rather than as an end result.

4. Environmental or situational variables provide the external dimension of individual development. Knowledge, understanding, skills, attitudes, values, and aspirations are the product of the interaction of these external variables with the internal variables that characterize the individual.
5. The developmental learning process moves from a beginning level of awareness and differentiation (*perceptualization*), to the next level of conceptualizing relationships and meanings (*conceptualization*), to the highest level of behavioral consistency and effectiveness by both internal and external evaluation (*generalization*).

The Career Conscious Individual Model's guidance curriculum has as its beginning point fifteen outcome-oriented goals. The goals are global statements of student outcomes around which the curriculum is constructed. The goals represent what students should possess in the way of knowledge, understandings, and skills as a result of their involvement in the guidance curriculum. These goals by domain are as follows:

Domain I: *Self-knowledge and Interpersonal Skills*

Goal A: Students will develop and incorporate an understanding of the unique personal characteristics and abilities of themselves and others.

Goal B: Students will develop and incorporate personal skills that will lead to satisfactory physical and mental health.

Goal C: Students will develop and incorporate an ability to assume responsibility for themselves and to manage their environment.

Goal D: Students will develop and incorporate the ability to maintain effective relationships with peers and adults.

Goal E: Students will develop and incorporate listening and expression skills that allow for involvement with others in problem-solving and helping relationships.

Domain II: *Life Roles, Settings, and Events*

Goal A: Students will develop and incorporate those skills that lead to an effective role as a learner.

Goal B:	Students will develop and incorporate an understanding of the legal and economic principles and practices that lead to responsible daily living.
Goal C:	Students will develop and incorporate an understanding of the interactive effects of life-styles, life roles, settings, and events.
Goal D:	Students will develop and incorporate an understanding of stereotypes and how stereotypes affect career identity.
Goal E:	Students will develop and incorporate the ability to express futuristic concerns and the ability to imagine themselves in these situations.
Domain III:	*Life Career Planning*
Goal A:	Students will develop and incorporate an understanding of producer rights and responsibilities.
Goal B:	Students will develop and incorporate an understanding of how attitudes and values affect decisions, actions, and life-styles.
Goal C:	Students will develop and incorporate an understanding of the decision-making process and how the decisions they make are influenced by previous decisions made by themselves and others.
Goal D:	Students will develop and incorporate the ability to generate decision-making alternatives, gather necessary information, and assess the risks and consequences of alternatives.
Goal E:	Students will develop and incorporate skill in clarifying values, expanding interests and capabilities, and evaluating progress toward goals.

Each of the fifteen goals is divided into a perceptual objective, a conceptual objective, and a generalization objective. This was done based on assumption 5, the assumption that stated that learning begins at the perceptual level, then moves through a conceptual phase, and ends at the generalization level. Individual learning moves from initial awareness to a state of behavior integration. Ordinarily, since the curriculum begins with perceptual objectives, these are emphasized at the elementary level. Conceptual objectives are middle or junior-high-school–focused, and generalization objectives are stressed at the senior-high-level. However, this placement of these objectives does vary somewhat by grade level across the fifteen goals (see Appendix).

These three objectives per goal are further divided into specific student competency statements by grade level. The student competency statements represent the specific behaviors (knowledge, skills, attitudes) that those involved in the program will exhibit or demonstrate. You will note that no mention is made of behavioral or performance objectives. The curriculum presented here moves directly to student competencies. One reason for this is to reduce the complexity of the curriculum design. The most important reason, however, is that a shift in the program is made almost immediately to students. Teachers and counselors have behavioral or performance objectives for their work; students have competencies. Thus we recommend the competency approach for the design of the guidance curriculum so that students can begin to assume responsibility for their own development.

To illustrate this format, we have selected Goal A from the Self-knowledge and Interpersonal Skills Domain. Objective 1, a perceptualization objective, has under it four competencies grades K–3. Objective 2, a conceptualization objective, follows with seven competencies grades 4–10. Finally, the generalization objective, objective 3, has under it two grades, grades 11 and 12. Example objectives and competencies for all of the fifteen goals for all grade levels appear in the Appendix.

I. *Self-knowledge and Interpersonal Skills*

A. Students will develop and incorporate an understanding of the unique personal characteristics and abilities of themselves and others.

1. Students will be aware of the unique personal characteristics of themselves and others.

   Students will

   a. describe their appearance and their favorite activities. (Kindergarten)

   b. recognize special or unusual characteristics about themselves. (first grade)

   c. recognize special or unusual characteristics about others. (second grade)

   d. describe themselves accurately to someone who does not know them. (third grade)

2. Students will demonstrate an understanding of the importance of unique personal characteristics and abilities in themselves and others.

   Students will

   a. analyze how people are different and how they have different skills and abilities. (fourth grade)
   b. specify those personal characteristics and abilities that they value. (fifth grade)
   c. analyze how characteristics and abilities change and how they can be expanded. (sixth grade)
   d. compare their characteristics and abilities with those of others and will accept the differences. (seventh grade)
   e. describe their present skills and predict future skills. (eighth grade)
   f. value their unique characteristics and abilities. (ninth grade)
   g. analyze how characteristics and abilities develop. (tenth grade)

3. Students will appreciate and encourage the unique personal characteristics and abilities of themselves and others.

   Students will

   a. specify which characteristics and abilities they appreciate most in themselves and others. (eleventh grade)
   b. appreciate their uniqueness and encourage that uniqueness. (twelfth grade)

As you read over the perceptualization, conceptualization, and generalization objectives and the competencies under each goal in the foregoing example and in the Appendix, you will note that they are developmental and sequential. The guide that was used in the writing of these objectives and competency statements was the taxonomy of Wellman and Moore (1975). Briefly stated it is as follows:

I. Domains

   A. Goals

      1. *Perceptualization Level*

         a. Student competencies concerned with environmental orientation
         b. Student competencies concerned with self-orientation

2. *Conceptualization Level*
    a. Student competencies concerned with directional tendencies
    b. Student competencies concerned with adaptive and adjustive behaviors
3. *Generalization Level*
    a. Student competencies concerned with accommodation
    b. Student competencies concerned with satisfaction
    c. Student competencies concerned with mastery

Perceptualization Level. Competencies at this level emphasize the acquisition of knowledge and skills, and attention to selected aspects of environment and self. The knowledge and skills most relevant are those needed by individuals in making appropriate life role decisions and in responding to the demands of school and the social environment. Attention is the first step toward the development and maturation of interests, attitudes, and values. Competencies at the perceptualization level reflect accuracy of perceptions, ability to differentiate, and elemental skills in performing functions appropriate to the individual's level of development. Competencies at this level are classified under two major categories, *environmental orientation* and *self-orientation.*

Competencies classified as environmental orientation emphasize the individual's awareness and acquisition of knowledge and skills needed to make life role decisions and to master the demands of life career settings and events. The competencies at this level are essentially cognitive in nature and have not necessarily been internalized to the extent that the individual attaches personal meaning to the acquired knowledge and skills. For example, individuals may acquire appropriate study skills and knowledge, but it does not necessarily follow that they will use these skills and knowledge in their study behavior. However, such knowledge and skills are considered to be prerequisites to behavior requiring them. Thus, the acquisition of knowledge and skills required to make growth-oriented decisions and to cope with environmental expectations is viewed as the first step in the development of individuals, regardless of whether subsequent implementation emerges. A primary and universally applicable goal of guidance is the development of knowledge and skills to enable individuals to understand and meet the expectations of their school and social environment and to recognize the values underlying social limits.

Competencies classified as self-orientation focus upon the development of accurate self-perceptions. One aspect of an accurate awareness of self is the knowledge of abilities, aptitudes, interests, and values that characterize individuals. An integral part of identity is individuals' ability to understand and accept the ways that they are alike and different from other individuals. Attention to life career decisions and demands relevant to immediate adjustment and future development is considered a prerequisite to an understanding of the relationships between self and environment. An awareness, and perhaps an understanding, of feelings and motivations is closely associated with self-evaluation of behavior, with the formation of attitudes and values, and with voluntary, rationally based modification of behavior. The goal of guidance at this level is to help individuals make accurate assessments of self so that they can relate realistically to their environment in their decisions and actions. The goal of guidance is individuals' development of self-awareness and differentiation that will enable appropriate decision making and mastery behavior in the roles, settings, and events of their lives.

Conceptualization Level. Individual competencies at the conceptualization level emphasize action based upon the relationships between perceptions of self and perceptions of environment. The types of action sought are categorized into personally meaningful growth decisions and adaptive and adjustive behavior. The general goal at this level of development is that individuals will (1) make appropriate choices, decisions, and plans that will move them toward personally satisfying and socially acceptable development; (2) take action necessary to progress within developmental plans; and (3) develop behavior to master their school and social environment as judged by peers, teachers, and parents. The two major classifications of conceptualization objectives are *directional tendencies* and *adaptive and adjustive behavior.*

The directional tendencies relate to movement of individuals toward socially desirable goals consistent with their potential for development. These competencies are indicators of directional tendencies as reflected in the choices, decisions, and plans that individuals are expected to make in ordering the course of their educational, occupational, and social growth. The acquisition of knowledge and skills covered by competencies at the perceptual level is a prerequisite to the pursuit of competencies in this category, although the need to make choices and decisions may provide the initial stimulus for considering perceptual competencies. For example, a ninth-grader may be required to make curricular choices that have a bearing upon post-high-school education and occupational aspirations. The need to make an immediate choice at this point may stimulate an examination of both environmental perceptions and self-perceptions as well as a careful analysis of the relationships between the two. To this extent, then, the interrelationship and interdependence of perceptual and conceptual competencies precludes the establishment of mutually exclusive categories. Furthermore, the concept of a developmental sequence suggests this type of interrelationship. Any choice that may determine the direction of future development is considered to represent a directional tendency on the part of individuals, and competencies related to such choices are so classified.

The expected emergence of increasingly stable interests and the strengthening and clarification of value patterns constitute additional indicators of directional tendencies. Persistent attention to particular persons, activities, or objects in the environment, to the exclusion of others (selective attention), is an indication of the development of interests through an evaluation of the relationships of self to differentiated aspects of the environment. Objectives that relate to value conceptualization, or the internalization of social values, complement interest development. Here individuals are expected to show increased consistency in giving priority to particular behavior that is valued personally and socially. In a sense, the maturation of interests represents the development of educational and occupational individuality, while the formation of value patterns represents the recognition of social values and the normative tolerances of behavior.

Competencies in these subcategories include consistency in the expression of interests and values and the manifestation of behavior compatible with the emerging interests and value patterns. For example, high school students may be expected to manifest increasing and persistent interests (measured or expressed) in particular persons, activities, and objects. They may be expected to develop a concept of self that is consistent with these interests and to place increasing importance, or value, on behaviors, such as educational achievement, that will lead to the development of related knowledges and skills and to the ultimate achievement of occupational aspirations. The directional tendency emphasis is upon achieving increased consistency and strength of interests and values over a period of time. The incidental or occasional expression of an immediate interest or value with little or no long-range impact upon the behavior of individuals should not be interpreted as an indication of a directional tendency.

The second major category of objectives at the conceptualization level includes those related to the application of self-environment concepts in coping with environmental presses and in the solution of problems arising from the interaction of individuals and their environment. Competencies in this area of functioning are designated as adaptive and adjustive behavior.

Adaptive behavior refers to the ability and skill of individuals in the management of their school and social environment (with normative tolerances) to satisfy self needs, to meet environmental demands, and to solve problems. There are two types. First, individuals may, within certain prescribed limits, control their environmental transactions by selection. For example, if they lack the appropriate social skills, they may avoid social transactions that demand dancing and choose those where existing abilities will gain the acceptance of the social group. Second, individuals may be able to modify their environment to meet their needs and certain external demands. For example, students who find sharing a room with a younger brother or sister disruptive to studying may be able to modify this situation by arranging to study elsewhere.

Adjustive behavior refers to the ability and flexibility of individuals in modifying their behavior to meet environmental demands and to solve problems. Such behavior modification may include the development of new abilities or skills, a change of attitudes, or a change in method of operation or approach to the demand situation. In the examples of adaptive behavior just mentioned, individuals might use adjustive behavior by learning to dance rather than avoiding dancing, and they might develop new study skills so they are able to study while sharing a room.

The basic competencies in this area are individuals' ability to demonstrate adaptive and adjustive behavior in dealing with school and social demands and in solving problems that restrict their ability to meet such demands. The competencies may be achieved by the application of existing abilities or by learning new ways of meeting demands.

Generalization Level. Competencies at the generalization level imply a high level of functioning that enables individuals to (1) accommodate environmental and cultural demands; (2) achieve personal satisfaction from environmental transactions; and (3) demonstrate competence through mastery of specific tasks and through the generalization of learned behavior, attitudes, and values to new situations. Behavior that characterizes the achievement of generalization-level competencies may be described as purposeful and effective by self or intrinsic standards and by societal or extrinsic criteria. Individuals should be able to demonstrate behavioral consistency, commitment to purpose, and autonomy in meeting educational, occupational, and social demands. This, then, is a person who is relatively independent and predictable.

Guidance competencies at this level are classified as *accommodation, satisfaction,* and *mastery.* The concept of sequential and positive progress implies a continuous process of internalization, including applicational transfer of behavior and a dynamic, rather than a static, condition in the achievement of goals. The achievement of generalization competencies may be interpreted as positive movement (at each level of development) toward the ideal model of effective persons (self and socially derived) without assuming that individuals will ever fully achieve the ideal.

Accommodation competencies relate to the consistent and enduring ability to solve problems and to cope with environmental demands with minimum conflict. Accommodation of the cultural and environmental demands requires that individuals make decisions and take action within established behavioral tolerances. The applicational transfer of adaptive and adjustive behavior, learned in other situations and under other circumstances, to new demand situations is inferred by the nature of the competencies classified in this category. The achievement of accommodation competencies can probably best be evaluated by the absence of, or the reduction of, unsatisfactory coping behavior. The wide range of acceptable behavior in many situations suggests that individuals who perform within that range have achieved the accommodation competencies for a particular demand situation, whereas if they are outside that range, they have not achieved these competencies. For example, a student is expected to attend class, to turn in class assignments, and to respect the property rights of others. If there is no record of excessive absences, failure to meet teacher assignment schedules, or violation of property rights, it may be assumed that the student is accommodating these demands with normative tolerances. In a sense, the objectives in this category represent the goal that individual behavior conform to certain limits of societal expectancy, whereas the other categories of generalization competencies tend to be more self-oriented. The achievement of accommodation competencies may provide evidence of inferences regarding the congruence of individual values with the values of one's culture. Caution should be exercised in drawing such inferences, however, because the individual may demonstrate relative harmony externally but have serious value conflicts that do not emerge in observable behavior.

Satisfaction competencies reflect the internal interpretation that individuals give to their environmental transactions. Individual interests and values serve as the criteria for evaluating the decisions made and the actions taken within the guidance domains. Although the evaluations of parents, peers, and authority figures may influence individuals' interpretations (satisfactions), these competencies become genuine only as they are achieved in congruence with the motivations and feelings of individuals. The description of satisfaction competencies consistent with guidance programming should include the internal (individuals' evaluation of affiliations, transactions, and adjustments) in terms of personal adequacy, expectations, and congruency with perceived ideal lifestyle. Expressed satisfaction, as well as behavioral manifestations from which satisfaction may be inferred, such as persistence, would seem to be appropriate criterion measures. Also, congruency between measured interests and voluntarily chosen career activities should be considered.

Mastery competencies include the more global aspects of achievement and generalization of attitudinal and behavioral modes. Longer-range goals, encompassing larger areas of achievement, are emphasized here rather than the numerous short-range achievements that may be required to reach a larger goal. For example, a young child becomes aware of task demands and different ways to meet them (perceptualization). At the conceptualization level task-oriented behaviors are developed and made meaningful. Generalization (mastery) competencies reflect the internalization of these behaviors so that tasks are approached and achieved to the satisfaction of self and social expectations. In the social area competencies relate to social responsibility and contributions of individuals with respect to social affiliations and interactions appropriate to their developmental level. All of the competencies in this category are framed in the context of self and social estimates of potential for achievement. Therefore, criteria for the estimation of achievement of mastery competencies should be in terms of congruency between independent behavioral action and expectations for action as derived from self and social sources. For example, a mastery competency in the educational area might be achieved by high school graduation by one individual, whereas graduate work at the university level might be the expected achievement level for another individual.

Individual Planning

Concern for individual development in a complex society has been a cornerstone of the guidance movement since the days of Frank Parsons. In recent years the concern for individual development has intensified as society has become evermore complex. This concern is manifested in many ways, but perhaps it is expressed most succinctly in a frequently used goal for guidance: helping individuals become the persons they are capable of becoming.

Casting guidance and counseling in a personal development, personal advocacy role is not new. Lortie (1965) suggested that one grouping of tasks counselors might accept as part of their role would be that of advocate. The emphasis would be on "helping individual students cope effectively with the sometimes impersonal juggernaut of the school. . . . Advocate counselors would probably emphasize such values as humanism [and] individualism."*

Building on this same theme, Cook (1971) urged that school counseling and guidance claim the role of student advocate, with

* D. C. Lortie, "Administrator, Advocate or Therapist? Alternatives for Professionalization in School Counseling," in *Guidance: An Examination*, ed. R. L. Mosher, R. F. Carle, and C. D. Kenas (New York: Harcourt Brace Jovanovich, Inc. 1965), p. 137.

the end result being the enhancement of students' development. Similarly, Howard Miller, president of the Los Angeles Board of Education, supported the need for guidance and counseling programs to attend to the individual development of students. Writing in the *Los Angeles Times*, Sunday, July 17, 1977, about needed next steps for the Los Angeles schools, he described what kind of school programs were needed. Among the critically important programs he described were "extensive counseling resources insuring personal direction and monitoring for each student" (Miller 1977).

It is clear from a review of the literature and an analysis of educational practices that concern does exist in education for individual development and that guidance is frequently seen as a vehicle to facilitate it. It is equally clear, however, that this concern is not always translated into organized guidance activities to carry out this mission systematically. This is true particularly in relationship to three critical and interlocking dimensions of individual planning–placement, follow-up, and follow-through (Wasil, 1974; McDaniels and Simutis, 1976). As a result, the individual-planning component, with its focus on these dimensions, was made an integral part of the total Career Conscious Individual Model.

To accomplish the purposes of this component of the model, activities and procedures are needed to assist students in continuously monitoring and understanding their growth and development in terms of their personal goals, values, abilities, aptitudes, and interests (competencies) so that they can take action on their next steps educationally and occupationally. This means that counselors and others with guidance responsibilities serve in the capacity of personal-development-and-placement specialists. Personalized, continuous contact and involvement with individuals are required.

To illustrate the nature and structure of this component, we chose an example that describes one way of organizing school staff to respond to the individual development of students. The example describes the idea of an advisory system in a high school setting. This approach is not new, but it did receive substantial attention in the 1970s. The entire issue of *NASSP Bulletin* (September 1977) was devoted to it. Before that, Hubel, Tillquist, Riedel, and Myrbach (1974) and Hubel (1976) wrote about the teacher-advisor system and how to put it into action. More recently, Johnson and Salmon (1979) described the nature and impact of an advisory system in one school district.

By choosing to illustrate the individual-planning component with the advisory system idea does not mean that we place less emphasis on placement, follow-up, and follow-through. On the contrary, without these critical program elements, a guidance program is not comprehensive. Space limitations precluded a full discussion of these elements, however. Also, there is substantial literature already available that describes in detail how to plan, implement, and evaluate these three elements. But to make sure we have made our point, remember, a comprehensive program is not complete without placement, follow-up, and follow-through activities.

The Individual Advisory System.[8] The Individual Advisory System (IAS) is based on the belief that satisfaction on the part of the faculty, students, and parents will result more easily if every student in school is able to relate personally, in a comfortable way, with at least one adult. In order for this one-to-one relationship to exist it is necessary to involve faculty members in a program that includes all students and their parents. Each teacher, counselor, administrator, and specialist acts as an advisor to a group of fifteen to twenty students. Thus, within these groups students are relating to one another as human beings sharing more than subject-matter concepts. The faculty members relate to one another as advisors sharing ideas about successfully dealing with their advisees. Parents relate to an individual in school who knows more about their child than grades earned in a particular class.

Program planning, parent contact, and personal development are the three main areas of the advisor's role. But advisors have other responsibilities to their advisees and to their school. In the following list are seven categories of advisor responsibility with the definition of each. Obviously, advisors will not cover every area with every advisee completely. Also in-service workshops are

[8] This section on the Individual Advisory System was written by Suzanne Fitzgerald Dunlap and Edna Erickson Bernhardt. It was written when they were research associates in the counseling and personnel services department, University of Missouri–Columbia. It was based on material developed as a result of an ESEA, Title II Project, Ferguson-Florissant School District, 1975. Reproduced by permission of the authors and McCluer North Senior High School, Florissant, Missouri.

needed to assist advisors in developing skills in each of the seven areas.

Program Planning: Any activity dealing with the act of choosing school courses, such as course selection, evaluation of course schedule, or tentative long-range educational planning.

Self-assessment: The analysis an advisee makes of his behavior, performance, or actions in an effort to strive for continuous self-improvement and understanding. All goal setting activities are included in this category.

School Offering Awareness: Any activity that contributes to an awareness of the school and its programs, philosophies, and actions.

Parent Relations/Conference: Those special activities designed to increase parent participation in the schooling process of their children and to ensure frequent positive contact among the advisor, student, and parent.

Feedback/Evaluation: Information which a school needs to hear, formally or informally, so that it can change itself to better suit the needs and desires of the people it serves. This category does *not* mean feedback to the student. It means feedback a student gives *to the school.*

Decision-Making Skills: The conscious application of a process to make decisions. Although decision making is woven into activities in many categories, it also is a distinct category to aid advisors in teaching the process.

Career Planning/Preparation: Activities to help students select and prepare for a career.

School/Community Issues: Activities concerned with the human aspects of individuals working together. Included are human development activities and group building. This area also includes any discussions needed about current school-wide issues that might arise during a school year, such as vandalism, a special decision the school needs to make, or any shared concern. (Hawkins and Cowles, 1975, pp. 7-9.)

Being an advisor makes it possible to give a few individual students personal care and attention. An advisor has fifteen to twenty advisees for all of their high school years, if this is mutually satisfactory. A student chooses an advisor in one of several ways. The choice may be based on curriculum or out-of-school interests of the advisor. The choice may be an advisor a student already knows. It may be that a group of students choose the same advisor so they can be together.

Advisors are given the permanent cumulative records of their advisees. This allows for easy access to student files during advisor-advisee meetings. In some schools folders are kept in an Advisement Center, which is also where advisor-advisee meetings may be held. Meetings are held during a regularly scheduled time. The scheduled advisement period usually takes priority over any other commitment students might have.

The Individual Advisory System acknowledges the need of students to have long-term, personal relationships with advisors. The Individual Advisory System provides the time and structure necessary for this involvement to occur. It is accurate to say, then, that the work of school staff is supported and enhanced by the Individual Advisory System.

The counselor's role in the high school has traditionally included at least a number, if not all, of the following functions:

---

1. Crisis-oriented counseling.
2. Administering and interpreting tests.
3. Educational or vocational guidance for seniors.
4. Scheduling and student registration.
5. Student record upkeep.

The counselor's role in the high school that uses the Individual Advisory System includes:

---

1. Providing advisors with backup support by dealing with crisis-oriented referrals.
2. Administering and interpreting tests along with the other advisors.
3. Helping advisors develop skills in providing students with educational or vocational information.
4. Scheduling and student registration of fifteen to twenty advisees.

5. Student record upkeep of fifteen to twenty advisees.
6. Developing out-of-school learning programs.
7. Helping advisors develop skills in active listening, group dynamics, parent conferencing, and conflict resolution.

When we compare the two role descriptions, it becomes evident that the number of responsibilities does not vary greatly. What is strikingly different is the manner in which each method meets the needs of students and uses the talents of counselors.

Traditionally, one counselor has been responsible for as many as four to five hundred students or more in a school. As a result, interaction between these students and the counselor has often been limited to brief encounters during registration times or other encounters during problematic times. Interaction between teacher and counselor has been limited, too, to faculty meetings, workshops, and brief consultations regarding problem students. The school program has simply not included the means for significant and consistent dialogue between counselors, faculty members, and students.

With the Individual Advisory System at work in the high school, communication barriers between faculty members, and between faculty and students, are lessened. Personal caring becomes a priority in the school where there is equal involvement on the part of every faculty member and every student. Counselors become members of a team whose overall role includes meeting such student needs as receiving personal attention, learning how to assess oneself and set goals, entering into meaningful dialogue with parents and teachers, getting to know an adult in the school, experiencing daily emotional growth, and developing decision-making skills.

At first glance it may appear to counselors that implementation of the Individual Advisory System would serve only to minimize their function in a school. If suddenly every other faculty member is called upon to advise a small group of students (help them plan their school programs, facilitate their emotional growth, and maintain contact with their parents), how, then, do counselors' special talents come into play?

Counselors functioning in a program of advisement serve as prime resources for both advisors and students. Freed from an abundance of paperwork and now integrated into the mainstream of the high school, they are finally able to use their talents and training. Counselors will continue to fulfill many of the traditional roles, such as crisis counseling, but the Individual Advisory System frees them to deliver a higher level of professional services. Very simply stated, an advisement program can enhance the work of counselors rather than jeopardize their position.

For some principals, student contact involves mainly disciplinary action with resulting negative feelings. The Individual Advisory System does not guarantee principals freedom from that task, but personalized contact and program planning will increase student motivation and involvement in the school. This is not to say that principals will never have discipline problems, but they should have fewer of them. IAS will help prevent these problems.

The Individual Advisory System needs principals to function as advisors. This means that principals will also have the opportunity for full participation in helping relationships. In addition, principals can serve as models for other advisors. As model advisors, they need to sustain participation with enthusiasm. In fact, principals help provide for long-term maintenance of the program.

Making changes is never easy. The rest of the staff will need support as they learn the role of advisor and make necessary adjustments and improvements. As advisors, principals develop an increased sensitivity to the problems of advisement. Their full participation demonstrates the value and importance of advisement. Seeing principals share equally the responsibilities of advisor will enhance the feeling of togetherness among the faculty. Everyone needs to feel trusted and cared for: counselors, teachers, administrators, and students. The environment should say to all within: You are trusted! Principals need to do all that they can do to create this feeling. It includes placing confidence in IAS, the staff, the students, and themselves.

The skills needed to be an effective advisor are not so different from those required to be a good teacher or principal or counselor. What is different is the one-to-one contact with a student outside of a curriculum-based situation. It is essential that advisors deal with their advisees in an honest and comfortable manner. In other words, advisors are most effective when they are themselves. Realization of this basic point will make the one-to-one contact easier to deal with. Once advisors reach this point, it is time to assess strengths and weaknesses and determine those areas where more skill development is needed.

Advisors should have some understanding of how to develop skills useful in attending to the personal growth and well-being of advisees, helping each advisee outline a satisfactory learning program, and maintaining communication with the advisee's parents. Advisor skills include career planning and preparation, college information, conflict resolution, decision making, interpreting test scores, parent conferencing, program planning, record keeping, school awareness, and self-assessment. It is not realistic to expect every advisor to master each of the listed skills. Advisors need to share their knowledge and help one another develop the skills. As in the case of advisor responsibilities, it is helpful if advisors present in-service workshops for one another.

It is important to keep the basic purpose of an individual advisory system in focus when organizing the program. It is possible for an advisory-type system to be oriented toward school-system efficiency rather than personalizing student learning. While the system should be reasonably efficient, an overconcern for efficiency can distort the purpose. It is possible to have an Individual Advisory System and not increase individualized student contact and involvement. A warmed-over homeroom approach, wherein students are only seen in a group and the goals are limited mainly to such administrative concerns as class scheduling, record keeping, and announcements, will probably offer little personal development. Although class scheduling may be one responsibility of advisors and group meetings may be an appropriate method for discussion and efficient processing, the primary focus is always on the facilitation of individual student development.

Responsive Services

Problems relating to academic learning, personal-identity issues, drugs, and peer and family relationships continue to be a part of the educational scene. As a result there is a continuing need for crisis counseling, diagnostic and remediation activities, and consultation and referral to be an ongoing part of a comprehensive guidance program. In addition, there is a continuing need for the guidance program to respond to the immediate information-seeking needs of students, parents, and teachers. The on-call responsive component of the Career Conscious Individual Model organizes guidance and counseling techniques and methods to respond to these concerns and needs as they occur. In addition, responsive activities are supportive of the guidance curriculum and individual development, placement, and follow-through activities.

Adjunct guidance staff—peers, paraprofessionals, volunteers—can aid counselors in carrying out on-call responsive activities. Peers can be involved in tutorial programs, orientation activities, ombudsman functions, and—with special training—cross-age counseling and leadership in informal dialog. Paraprofessionals and volunteers can provide assistance in such areas as placement, follow-up, and community-school-home liaison activities.

Rather than illustrating this program component by describing example activities, we chose instead to describe the physical facilities for the total guidance program. We did this because there is already available an abundance of literature that describes in detail techniques and procedures to be used in carrying out activities in the responsive-services component. Resources for this component also abound and are well known, so they are not mentioned here.

We chose to describe guidance facilities because not much attention has been given to this topic in the past. What attention has been given has been done based on traditional ways of organizing guidance. To make the responsive-services component function effectively, to provide direct and immediate responses to individual needs, and support the other program components, a new way of organizing guidance facilities is needed. They have traditionally consisted of an office or suite of offices designed primarily to provide one-to-one counseling assistance. Such an arrangement frequently included a reception or waiting area that served as a browsing room where students had limited access to some displays or files of educational and occupational information. The need for individual offices is obvious because of the continuing need to carry on individual counseling sessions. There is also a need, however, to open up guidance facilities, to make guidance facilities more accessible and usable.

One approach to make guidance facilities more accessible and usable is to reorganize traditional space into a Career Center. Other titles used to describe these centers include Career Planning and Placement Center, and Career Resource Center. Pick the title that suits your situation. We have chosen to call these reorganized facilities the Career Center.

The Career Center. A comprehensive Career Center can bring together available guidance information and exploration resources and make them easily accessible to students. The center can be used for such activities as research, planning, self-exploration, and group sessions. Students can gain assistance in such areas as occupational planning, job entry and placement, financial aid information, and postsecondary educational opportunities.

Although the center is available for use by school staff and community members, it should be student-centered, and many of the center activities should be student-planned as well as student-directed. At the same time, the center is a valuable resource for teachers in their program planning and implementation. Employers, too, will find the center useful when seeking part-time or full-time workers. Viewed in this way, the impact of the center on school and community can be substantial.

If community members and parents are involved in the planning and implementation of the center, their interest could provide an impetus for the involvement of other community members. When parents and other community members become involved in programs such as the Career Center, they gain firsthand experience with the educational process. Through these experiences new support for education can grow.

The Career Center should be furnished in as comfortable a way as possible for all users. Provision should be made for group as well as individual activities. Coordination of the operation of the Career Center should be the responsibility of the guidance staff. All school staff should be involved, however. It may be necessary to employ at least one paid paraprofessional to ensure that clerical tasks are carried out in a consistent and ongoing manner. The majority of the staff could be comprised of volunteers—both community members and students. This provides an excellent opportunity for the involvement of these persons in the school program. Teachers and administrators should also be encouraged to participate. Their participation could provide students with the opportunity to get to know them on a more casual basis.

The activities of the Career Center will be as varied as the people involved in its operation. Many guidance activities are ideally suited for processing through the center. Some of these are career exploration groups, the providing of occupational information, the providing of educational information, peer counseling, an individual advisement program, a community resource survey, and a work exploration program.

Placement should also be an integral part of the activities offered by the Career Center. Through placement, counselors and teachers can assist students in carrying out educational and occupational plans suitable to their interests, skills, abilities, and training.

In order to have successful placement activities in a center, there are several necessary elements: outreach to students, copartnership with fellow educators and employers, and meaningful follow-ups on which to build and adapt curriculum. The following activities are suggestions for providing these elements:

---

1. Outreach activities involving such things as announcements concerning the center posted on bulletin boards or announced over intercom systems or local radio or television programs; assemblies arranged for the job placement coordinator to speak to students and discuss the current job market, preparatory steps for job hunting, and the initial interview; student referrals to the center by counselors, teachers, and administrators; counseling and placement assistance for students who withdraw from school.
2. Student assessment, including both formal and informal testing procedures to help students become aware of their goals, interests, abilities, achievements, and personal characteristics.
3. Community surveys involving personal contact with individuals in industry, business, professions, private agencies, and employment offices to provide up-to-date information for the community resource file in the center. This file can be used by educators and community members to identify possible locations for work exploration, on-site visits, classroom visitors, and job placement. Other useful resources to be placed in the file can be found through the local Chamber of Commerce, civic organizations, personnel associations, business associations, parents, and teachers.
4. Structured experiences preparing students for educational and occupational placement, such as filling out applications, interviews and telephone tips, testing (college entrance examinations, civil service

testing, employment examinations), appearance, salary, budgeting, benefits, and unions.

5. Activities assisting instructional staff to relate their subject matter to the world of work and assisting youth in the development of appropriate work attitudes and understandings.
6. Evaluation activities, including follow-up studies of former students.

Although the center as described here is planned specifically for the secondary school, the concept may be modified and made applicable for either elementary or postsecondary education. At the elementary level the center might serve as the coordination center for classroom activities. Salmon and Selig (1976) discussed the center concept in the elementary school and provided a number of examples of activities that could be included in such a center. The postsecondary focus could be much the same as that of the secondary program. Reardon and Burck (1977) described procedures and methods to establish such centers in postsecondary institutions, as did Beaumont, Cooper, and Stockard (1978).

Additional references on how to develop career centers and how to train counselors and other personnel to operate them are available. Axelrod, Drier, Kimmel, and Sechler (1977) put together a handbook on how to develop and operate centers. Jacobson (1974a, 1974b) developed two filmstrips—one for junior high schools and one for senior high schools—that provide a visual display of centers and how they operate. Training ideas for center staff can be obtained from material developed by Johnson (1976).

The number of Career Centers has increased rapidly since the early 1970s. Centers have met with increasing acceptance by counselors, teachers, students, administrators, parents, and members of the community. Jacobson (1978), put it this way:

> The rapid increase in the establishment of career resource centers in the last eight years has had and will continue to have a profound effect on all of us who are associated with guidance and counseling programs. To assume otherwise is to ignore an imperative for guidance that has a possibility of achieving your wildest expectations for guidance program acceptance, effectiveness, and growth, whether you are a counselor, a

guidance director, a counselor educator, or a state supervisor of guidance.[9]

System Support

The administration and management of a comprehensive guidance program requires an ongoing support system. That is why it is a major program component. Unfortunately, however, it is an aspect of a comprehensive program that is often overlooked or, if it is attended to, only minimally appreciated. And yet, the system-support component is as important as the other three components. Why? Because without continuing support the other three components of the guidance program may be less than effective.

Activities included in this program category are by definition activities that support and enhance activities in the other three program components. That is not to say that these activities do not stand alone. They can and often do. But for the most part, they undergird activities in the other three components.

For our purposes, activities in the system-support category include but are not limited to the following:

---

1. Staff development
    - Assessment and planning
    - Implementation
2. Orientation
    - Between schools
    - Students, parents, staff, community
3. Testing
    - Organization and planning
    - Giving and reporting
4. Community relations
    - Linkage with labor, business, and industry

[9] T. J. Jacobson, "Career Resource Centers," in *New Imperatives for Guidance*, eds. G. R. Walz and L. Benjamin (Ann Arbor, MI.: ERIC Counseling and Personnel Services Clearinghouse 1978), p. 416.

Dissemination of information

5. Materials development
   Student curriculum materials
   Staff support materials
6. Guidance program review
   Needs assessment (current status and desired outcome)
   Evaluation (outcome and process)
7. Parent education
   Materials dissemination
   Workshops, seminars, meetings

Staff Development. To illustrate a system-support activity in more detail, we have chosen staff development. Staff development is crucial to the success of the activities in the other three program components. Staff development is for members of the guidance staff as well as for the teaching staff. Staff development is an ongoing activity.

To understand all that is involved in planning and conducting staff development activities, it is first necessary to understand the essential elements needed for staff development to take place. The table on p. 97 outlines these elements and conditions.

Table 3.1 provides a description of the phases of learning individuals go through as they relate the essential elements of staff development. For example, for a new skill or technique to be learned, activities need to be constructed first to make an individual aware and then to allow some experimentation in trying out the skill or technique so that he or she can become competent in that skill or technique. It is important that staff development activities include all the elements described (knowledge to be learned, skill or technique needed, attitudes or values required, and application stages) and attend to the appropriate learning phases to be effective. If only knowledge to be learned is dwelt upon and staff are not provided with the skills or techniques to apply the knowledge, the impact of the activities will be lessened considerably.

Once the essential elements and learning phases of staff development are understood, the first step in conducting staff development activities is to assess the competencies of staff and their

TABLE 3.1

Staff Development Model

	Essential Elements			
Learning Phases	*Knowledge to Be Learned*	*Skills or Techniques Needed*	*Attitudes or Values Required*	*Application Stages*
Attending	Perceptualization	Awareness	Identification	Demonstration
Relating	Conceptualization	Experimentation	Clarification	Simulation
Internalizing	Generalization	Competency	Affirmation	Performance

perceptions of the need for further training. Based on staff needs assessment, the next step is to use the staff development model to bring appropriate activities together. For example, if one or more of the goals of the guidance program is for individuals to develop human relationship skills, it will be necessary to increase staff competencies in human relations skills. Here is an example of how the model works:

Human Relationship Skills	
Knowledge to be learned	Carkhuff-helping-model facilitative conditions
Skills or Techniques Needed	Role playing Peer counseling Attending Disclosing
Attitudes or values required	Involvement Trust
Application stages	Teacher Effectiveness Training Group counseling

As you can see, the model outlines what is to be learned, the skills and attitudes or values involved, and the procedures to be used to accomplish the task. An additional example follows:

Employability	
Knowledge to be learned	Job-seeking skills Occupational information
Skills or techniques needed	Using *Occupational Outlook Handbook* Role-playing interviews Cooperation Task completion
Attitudes or values required	Accommodation Interdependency
Application stages	Mini-course-job seeking Visits to local industry

Once staff needs have been assessed and activities have been chosen to meet those needs, the next step is to locate resources to carry out the activities. Most school districts have access to a variety of resources to support staff development activities. Here are a few examples:

---

1. Workshops by local staff.
2. University courses.
3. University-sponsored workshops.
4. An individual or team sent away to a source (workshop) to bring back information and skill to share with others.
5. An individual or team sent to observe model projects.
6. Creation of a consortium of local school districts to use one another's expertise.
7. Use of an outside consultant over a period of time.
8. State Department of Education workshops.
9. Self-instructional packages.
10. Instructional television.

REFERENCES

Axelrod, V., H. N. Drier, K. S. Kimmel, and J. Sechler. *Career Resource Centers.* Columbus, Ohio: National Center for Research in Vocational Education, 1977.

Bailey, L. J., *Career and Vocational Education in the 1980s: Toward a Process Approach.* Carbondale, Ill.: Southern Illinois University, 1976.

Beaumont, A. G., A. C. Cooper, and R. H. Stockard, *A Model Career Counseling and Placement Program.* Bethlehem, Pa.: College Placement Services, Inc., 1978.

Bloom, B.S., ed. *Taxonomy of Educational Objectives: The Classification of Educational Goals. Handbook 1. Cognitive Domain.* New York: David McKay Co., Inc., 1956.

Borow, H., ed. *Career Guidance for a New Age.* Boston: Houghton Mifflin Company, 1973.

Cole, H. P., *Approaches to the Logical Validation of Career Development Curricular Paradigms.* Carbondale, Ill.: Career Development for Children Project, 1973.

———, *Process Education.* Englewood Cliffs, N.J.: Educational Technology Publications, 1972.

Cook, D. R., ed. *Guidance for Education in Revolution.* Boston: Allyn & Bacon, Inc., 1971.

Dudley, G. A., and D. V. Tiedeman, *Career Development: Exploration and Commitment.* Muncie, Ind.: Accelerated Development, Inc., 1977.

Eisenberg, S., "Exploring the Future: A Counseling-Curriculum Project," *Personnel and Guidance Journal*, 52 (1974), 527–33.

Goldhammer, K., "Career Education: An Humane Perspective on the Functions of Education," *Journal of Career Education*, 2 (1975), 21–26.

Goldhammer, K., and R. E. Taylor, *Career Education: Perspective and Promise.* Columbus, Ohio: Charles E. Merrill Publishing Company, 1972.

Gysbers, N. C., and E. J. Moore, "Career Development in the Schools," in *Contemporary Concepts in Vocational Education*, ed. G. F. Law. Washington, D.C.: American Vocational Association, 1971.

———, *Career Guidance, Counseling and Placement: Elements of an Illustrative Program Guide.* Columbia, Mo.: University of Missouri, 1974.

Hansen, L. S., "A Model for Career Development through Curriculum," *Personnel and Guidance Journal*, 51 (1972), 243–50.

Hansen, L. S., and W. W. Tennyson, "A Career Management Model for Counselor Involvement," *Personnel and Guidance Journal*, 53 (1975), 638–45.

Hawkins, M. L., and R. J. Cowles, *Just a Little Care.* Florissant, Mo.: ESEA, Title III Project, Ferguson-Florissant School District, 1975.

Herr, E. L., and S. H. Cramer, *Vocational Guidance and Career Development in the Schools: Toward a Systems Approach.* Boston: Houghton Mifflin Company, 1972.

Hubel, K. H., *The Teacher-Advisor System in Action.* Dubuque, Iowa: Kendall-Hunt Publishing Company, 1976.

Hubel, K. H., P. F. Tillquist, R. G. Riedel, and C. M. Myrbach, *The Teacher-Advisor System.* Dubuque, Iowa: Kendall-Hunt Publishing Company, 1974.

Jacobson, T. J., "Career Resource Centers," in *New Imperatives for Guidance*, ed. G. R. Walz, and L. Benjamin. Ann Arbor, Mich.: ERIC Counseling and Personnel Services Clearinghouse, 1978.

———, *High School Career Center.* Author, 5945 Highgate Court, La Mesa, Calif. 92041, 1974a (filmstrip with cassette).

———, *Junior High School Career Center.* Author, 5945 Highgate Court, La Mesa, Calif. 92041, 1974b (filmstrip with cassette).

Johnson, C., *Developing Facility Maintenance Competencies for Career Resource Technicians.* Palo Alto, Calif.: American Institutes for Research, National Consortium on Competency-based Staff Development, 1976.

Johnson, R. L., and S. J. Salmon, "Caring and Counseling–Shared Task in Advisement Schools," *Personnel and Guidance Journal*, 57 (1979), 474-77.

Jones, G. B., J. A. Hamilton, L. H. Ganschow, C. B. Helliwell, and J. M. Wolff, *Planning, Developing and Field Testing Career Guidance Programs*. Palo Alto, Calif.: American Institutes for Research, 1972.

Krathwohl, D. R., B. S. Bloom, and B. B. Masia, *Taxonomy of Educational Objectives: The Classification of Educational Goals. Handbook 2. Affective Domain*. New York: David McKay Co., Inc., 1964.

Lortie, D. C., "Administrator, Advocate or Therapist? Alternatives for Professionalization in School Counseling," in *Guidance: An Examination*, ed. R. L. Mosher, R. F. Carle, and C. C. Kehas. New York: Harcourt Brace Jovanovich Inc. 1965.

McDaniels, C. O., and C. L. Simutis, "The Placement Service," in *School Guidance Services*, ed. T. H. Hobenshil and J. H. Miles. Dubuque, Iowa: Kendall-Hunt Publishing Company, 1976.

Mesa Guidance and Counseling Department, *Toward Accountability*. Mesa, Ariz.: Mesa Public Schools, 1973.

Miller, H., "Which Way Next for L.A. Schools?" *Los Angeles Times*, July 17, 1977, Part VIII, 5.

Prediger, D. J., J. D. Roth, and R. J. Noeth, *Nationwide Study of Student Career Development: Summary of Results*. Iowa City, Iowa: The American College Testing Program, 1973.

Reardon, R. C., and H. D. Burck, eds. *Facilitating Career Development: Strategies for Counselors*. Springfield, Ill.: Charles C Thomas, Publisher, 1977.

Reich, C. A., *The Greening of America*. New York: Bantam Books, Inc., 1971.

Rice, J. P., "Cooperative Guidance and Instructional Programs for Leadership Preparation," *Personnel and Guidance Journal*, 44 (1966), 967-73.

Richardson, H. D., and M. Baron, *Developmental Counseling in Education*. Boston: Houghton Mifflin Company, 1975.

Salmon, S. J., and M. R. Selig, "The Guidance Learning Center," *Elementary School Guidance and Counseling*, 10 (1976), 260-67.

Sanderson, B., and C. Helliwell, *Career Development Theory, Module 1: Developing Comprehensive Career Guidance Programs*. Palo Alto, Calif.: American Institutes for Research, 1975.

Super, D. E., "Career Education and Career Guidance for the Life Span and for Life Roles," *Journal of Career Education*, 2 (1975), 27-42.

———, *Career Education and the Meanings of Work*. Washington, D.C.: U.S. Government Printing Office, 1976.

Swan, R. J., "The Counselor and the Curriculum," *Personnel and Guidance Journal*, 44 (1966), 689-93.

Tennyson, W. W., L. S. Hansen, M. K. Klaurens, and M. B. Antholz, *Educating for Career Development.* St. Paul, Minn.: Minnesota Department of Education, 1975.

Wasil, R., "Job Placement: Keystone of Career Development," *American Vocational Journal,* 49 (1974), 32.

Wellman, F. E., and E. J. Moore, *Pupil Personnel Services: A Handbook for Program Development and Evaluation.* Washington, D.C.: U.S. Department of Health, Education and Welfare, 1975.

Chapter 4

An Improved Guidance Program: How to Get There from Where You Are

Now that you have identified a developmental program, the next step is to compare it with your current program. As we stated in Chapter 3, the model serves as a template to lay over the top of your current program so that similarities and differences can be seen. After the comparison you can use the processes of adopting, adapting, and creating in order to organize your improved guidance program (see Figure 4.1).

ORGANIZING YOUR IMPROVED PROGRAM

As your current program is being compared with a developmental program model, decisions are required concerning what the nature and structure of your improved guidance program should be. What should be adopted directly from the developmental program model? What current program components should be maintained? What program components not presently available in the program model need to be created in order to fill gaps in the current program? Your answers to these questions will provide answers to what the nature and structure of your improved guidance progam will be.

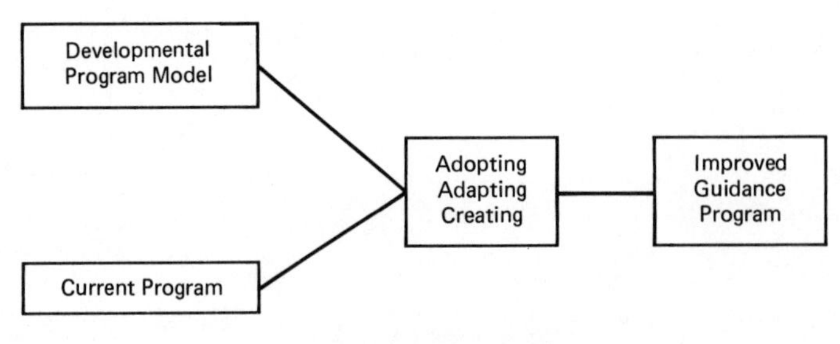

FIGURE 4.1.

Select Your Program Structure

One of the first decisions needed at this phase of the program improvement process is a decision concerning overall program structure. What should the structure be? What should the structure look like? Traditionally, as you will recall, a frequent response to these questions has been to use a structure known as guidance services. Over the years these services have included orientation, assessment, information, counseling, placement, and follow-up activities.

We recommended that in place of this structure, a new structure be adopted, a structure more in keeping with the developmental guidance perspective. We recommend that the components described in Chapter 3 be used as the organizers for your guidance program structure, K–12. Local and state needs will dictate the specific content to be included in each program component as well as how such content will be organized. The suggested program components are

---

Definition. The program definition identifies the centrality of guidance within the educational process and delineates in outcome terms the competencies individuals will possess as a result of their involvement in the program.

Rationale. The rationale discusses the importance of guidance as an equal partner in the educational process and provides reasons why individuals in our society need to acquire the competencies that will accrue to them as a result of their involvement in a comprehensive, developmental guidance program.

Assumptions. Assumptions are principles that shape and guide the program.

Guidance curriculum. The guidance curriculum contains the majority of guidance activities K–12. The curriculum contains goals, lists of competencies to be developed by students, and activities to assist students to achieve the competencies. The curriculum is organized by grade level and sequenced grades K–12. It is designed to serve all students.

Individual planning. Included in this component are guidance activities that assist individuals to understand and monitor their growth and development and to take action on their next steps educationally or occupationally with placement and follow-through assistance.

Responsive services. This component includes such activities as crisis-personal counseling, information giving, and consulting with staff and parents.

System support. Included in this component are activities necessary to support the activities in the other three components. Such activities as staff development, community resource development, and student assessment are included.

Decide on Student Competencies

Once you have selected your overall program structure, the next step is to decide on the student competencies for which the guidance program will take responsibility. What knowledge will students gain and what skills will students develop as a result of their involvement with the guidance program? For help in answering these questions, go to the results of the current program assessment you have already completed. The second step in what process identified the intended student competencies resulting from guidance activities K–12 in the current program. Begin with that list. Compare it with lists generated from the goals of your school district, from state Department of Education goals, or from local community goals. Some school districts and some state departments of education have developed competency lists to be used as a part of their graduation requirements. Use such lists in the comparison process. Also review the list of competencies presented in the Appendix. Then decide on the K–12 list to be used in your program. The preliminary work of assembling this list can be done by a work group. The final approval of this list, however, should be given by the total guidance staff as well as by key members of the school staff, administration, students, and the community. Use your school-community advisory committee to assist you in the process.

We recommend that in your organization of the K–12 competency list you use the following format. First, identify the broad categories into which the competencies fit. We call these categories domains. Next, establish overall goals for the program. Then organize the competencies around the goals. Or, the starting point can be competencies, then establish the goals, and finally identify the domains.

We stress the need to be parsimonious when it comes to the numbers of domains, goals, and competencies used as the basis for your improved program. The model presented in Chapter 3 used three domains and five goals per domain. This means that there were fifteen goals for the overall program model. For each grade level there were thirteen competencies, or 195 competencies K–12. We suggest that you not exceed these numbers, since longer lists are difficult to manage effectively. (See the Appendix for the list of 195 competencies.)

Choose Guidance Activities and Resources

Gelatt (1975) outlined four steps in selecting guidance activities and resources. First, he suggested that all imaginable activities and resources be surveyed. Next, all the possible activities and resources are listed. Then a list of desirable activities and resources is drawn up, and, finally, choices are made of the preferred activities and resources. To assist individuals with the first step, Gelatt (1975) identified two major groupings of activities and resources: instructional and guidance procedures and direct-indirect interventions. He provided the following examples in each of the groupings: (p. 19–22)

I. Instructional and guidance procedures
   A. Reading printed materials
      1. Narrative materials
      2. Programmed materials
      3. Cartoon booklets
      4. Kits

B. Observing
   1. Live demonstration
      a. Peer student models
      b. Cross-age models
   2. Live dramatizations
   3. Films
   4. Filmstrips
   5. Slides
   6. Videotapes
   7. Any one or all of the foregoing observational media followed by guided practice, supervised either by the models or by counseling personnel

C. Listening
   1. Radio
   2. Sound recordings
      a. Records
      b. Audiotapes

D. Interacting individually or in groups with
   1. Counseling personnel
   2. Community resource persons

E. Practicing behavior under simulated conditions
   1. Simulation games
   2. Simulated work
   3. Role playing
   4. Behavioral rehearsal

F. Gathering personal assessment information
   1. Responding to instruments measuring personal characteristics
   2. Collecting information from other people
   3. Self-assessment activities

G. Participating in computer-supported programs

H. Using on-line computer technology

II. Direct-indirect intervention

A. Direct interventions—learning activities employed directly with students
   1. Developmental activities
      a. *Orientation-in:* Orienting students for a new school level, a new educational system (for example, individualized

education), an innovative guidance program, or a new specific school setting

b. *Personal assessment*: Helping students understand and develop their own abilities, interests, physical attributes, personal and social behaviors and values, and preferences related to available career opportunities in six areas

c. *Personal-choice opportunities*: Assisting students to consider choices in each of the six areas of behavior—the career need areas

d. *Personal problem-solving skills*: Enabling students to make wise decisions and plans and to use the information gained in personal assessment and personal choice opportunities.

e. *Formulating and pursuing personal goals*: Assisting each student to formulate and to pursue his goals and his plans for achieving these goals in each of the six areas of behavior.

2. Prescriptive activities

   a. *Within-school learning experiences*: Working with one or more students having learning, social, or personal problems in the six areas of behavior

   b. *Orientation-out prescribed learning experiences*: Providing prescribed learning experiences for students at critical times—for example, beginning a new job, dropping out without specific plans, enlisting in the military

B. Indirect interventions—services provided on behalf of students

1. Interventions implemented through providing assistance to assess and possibly to modify

   a. Aspects of the educational setting and system

   b. School personnel

   c. Home and neighborhood factors

   d. Community resources (for example, health, social, and welfare agencies; business and industries)

2. Interventions implemented through

   a. Guidance-related research and evaluation, experimentally controlled studies of guidance and counseling materials and procedures, follow-up studies, and analyses of changes in the characteristics and needs of the student population

In step 2 of Gelatt's outline—listing all possible activities and resources—the task is to narrow down the list of all imaginable activities and resources to those which seem possible, to those which could be carried out. The key point in step 2 "is *not* to eliminate activities and resources you could do but may not want to do" (Gelatt 1975, p. 23). Next, in step 3, the task is to list all desirable activities and resources. Here is where local constraints and personal values come into play. Possible activities and resources can be reviewed in light of such things as

---

Magnitude: How much of the total school population or program will be affected?

Complexity: How many other changes will it incur?

Convenience: Can it be developed and operated locally, or will it require outside consultants?

Flexibility: How rigid must the method be followed in order to be successful?

Distinctiveness: Is it new and different?

Interaction with other programs: Does it stand alone or require other programs to be involved for success?

Readiness: Can it be applied immediately?

Cost: What are initial costs and future funding needs?

Content: Is the content an innovation or a redo of an old method? Gelatt 1975, p. 26)

Keeping Gelatt's suggestions in mind, the next task in the organization of your program is actually to choose the guidance activities and resources that will assist students to develop the competencies that have been decided upon. To complete this task we again suggest that you use the results of the current program assessment already completed. Take those current activities and those current resources you feel are in keeping with the improved program you are designing—those that have the potential of assisting students to develop the chosen competencies—and place them in the appropriate grade-level grouping.

In addition, we suggest you carefully examine the listing of activities and resources identified by Gelatt for possible new activities and resources (or perhaps ways of rewording current activities and resources) to add to those from the current program. There are many other such lists you also may wish to examine for possible ideas. Check with a local university or your state Department of Education for possible additional lists. Also keep in mind that there will probably not be a one-on-one match between a specific activity or resource and a specific competency. Often a single guidance activity or resource may result in the achievement of a number of student competencies.

To illustrate how an activity may be written and to show its relationship to the goals and competencies of the guidance curriculum suggested in Chapter 3, here are two examples. Both examples are from the Self-knowledge and Interpersonal Skills Domain. Both also relate to Goal A; one is for the fifth grade and one is for the tenth grade. You may wish to consider this format for your own activity descriptions. Note that each activity is one page in length and follows a straightforward outline.

Domain I
Goal A
Grade 5
Performance indicator 2

COAT OF ARMS

Domain I:	Self-knowledge and Interpersonal Skills
Goal A:	Students will develop and incorporate an understanding of the unique personal characteristics and abilities of themselves and others.
Competency:	Students will specify those personal characteristics and abilities that they value.
Performance indicator:	2–identify a variety of things that they value.

Activity objective

The student will identify six things he or she values.

Materials: Coat of arms, pencil.

Time: One session of approximately 30 minutes.

Directions: Have each student fill in an outline of a shield with the following for a personal Coat of Arms:

Box 1: Draw a symbol to represent your greatest success.

Box 2: Draw a symbol to represent your family's greatest success.

Box 3: Draw a symbol to show a place you dream about.

Box 4: Draw symbols to show two things at which you are good.

Box 5: Draw a picture to show what you would do if you have one year to live and would be a success at anything you do.

Box 6: On one-half of the box write two words you would like used to describe you; on the other half write two words you would *not* like used to describe you.

Domain I
Goal A
Grade 10
Performance indicators 1, 2

WHY I'M ME

Domain I: Self-knowledge and Interpersonal Skills

Goal A: The students will develop and incorporate an understanding of the unique personal characteristics and abilities of themselves and others.

Competency: Students will analyze how characteristics and abilities develop.

Performance indicators:

1–Explain the influence that genetic factors can have on characteristics and abilities.

2–Give examples of genetic and environmental factors that influenced the development of their own characteristics and abilities.

Activity objective

The student will list two genetic and three environmental factors that influence his or her life.

Materials: Paper, pencil.

Time: One class period of approximately 1 hour.

Directions: To insure that each student has an understanding of the basics, begin class with a brief discussion of genetics and what characteristics we may inherit.

Discuss the different environments of the country and of your state and how where we live influences our life-styles, career choice, attitudes, and so on. (Example: rushed life of those who live in New York City, keeping up with popular fads in California, work ethic of rural areas, and so on.)

Have each student make a list of two genetic and three environmental factors that influenced the development of their own characteristics and abilities.

If time permits, the students may get in small groups for discussion. Students should not be required to share what they have written if they choose not to.

Establish a Budget

In the assessment of your current program, one of the tasks was to review the financial resources available to you. In your review you may have found that the financial resources were described in terms of a well-defined budget in which all of the money spent on guidance including salaries was part of the budget. Or you may have found that only money for such items as testing materials and a few other guidance resources was included. Or, perhaps there was no budget at all. Within the budget policy guidelines for your school, your task is to establish a budget. As you are doing so, consider such major categories as

---

1. Personnel
    a. Counselors
    b. Secretarial-Clerical
    c. Fringe benefits
2. Materials and supplies
    a. Office supplies
    b. Testing materials
    c. Reference books
    d. Career literature
3. Equipment and maintenance
4. Professional development
    a. Meetings
    b. Consultants
5. Travel
6. Communication
7. Research and evaluation

A majority of the funds for your program will probably come from local sources. There are, however, funds from other sources of which you need to be aware. Federal legislation is one source. Currently, federal legislation dealing with employment, training, and education offer the greatest potential for funds. For example, the Comprehensive Employment Training Act Amendments of 1978, Title IV–Youth Programs, provides funding through prime sponsors for schools to conduct programs to serve in-school youth. Contact the prime sponsor in your area or your state Department of Education for more information.

The vast majority of federal funds for education are distributed through your state Department of Education. Currently, some funds for guidance are available from the Career Education Incentive Act (P.L. 95-207) and the Educational Amendments for 1978 (P.L. 95-561) through Title IV, Part B. A new part, Part D (guidance, counseling, and testing), was established for guidance under this legislation, but no funds were appropriated. If and when funds are appropriated, guidance funding will shift from Part B to Part D. In addition to these funds, funds also are available for guidance through vocational education legislation. Presently, such money is authorized under the Educational Amendments of 1976, Title II (P.L. 94-482). Again check with your state Department of Education for further information.

Keep in mind that most of these sources will not fund an entire program. They do, however, provide dollars for such aspects of a program as establishing a Career Center or maintaining a testing program. Also, keep in mind that most of the time access to these funds is through a proposal or the development of a written plan.

Some state legislatures have appropriated funds for guidance. For example, in 1978, the Oklahoma legislature appropriated funds for elementary school guidance programs. Grants were made to eligible schools provided that the programs met certain standards. Check your state for the possibility of such funding. Finally, don't overlook the possibility of contributors from such local sources as service clubs. Often club members are looking for projects. They may be interested in purchasing resource materials for a Career Center, sponsoring a Career Day, or providing for a specific in-service activity.

Modify Guidance Facilities

Now that you have selected your program structure and the student competencies for which you will assume responsibility, the next task is to review the physical facilities for the program. The reason that physical facilities are so important is that they often provide students with their first and sometimes lasting impression of the guidance program. Staff and parents also gain impressions about the guidance program and the work of the counselors through the physical facilities of the program. If a guidance office has many filing cabinets prominently displayed containing students records and in addition has the master schedule on the wall, it doesn't take long to know what the counselors do. If, on the other hand, the office space is organized as a Career Center, then another message is conveyed.

The question is, What impression do you want your program to convey? Your answer to this question will be answered in part by the program structure you selected and the student competencies for which the program will assume responsibility. Our bias suggests the use of a Career Resource Center approach to organizing guidance program facilities. In Chapter 3 you will recall there was discussion of Career Centers as well as identification of several excellent references that provide in-depth information about how to set up and operate such centers.

At one time space was at a premium in most schools. Counselors were fortunate to have an office. Although this is still true in some schools, most have more space available due to smaller enrollments. The best space is that nearest the place where student traffic is heaviest.

IMPLEMENTING YOUR IMPROVED PROGRAM

Once you have decided on the organization of your improved guidance program, the next step is to begin the implementation process. Ordinarily this process may take up to three years to unfold. A major reason for this is the time involved in beginning new activities and in shifting responsibility for previously assigned activities.

Assess Perceived Student Needs

To begin the implementation process we recommend that you assess the perceived needs of students. Some program planners do this task first, before any other step is taken in the planning process. While it can be done then, we don't believe the results will be as useful as the results of one done during this phase of the program improvement process.

A major reason we recommend waiting until now is that it isn't until now that you have identified the student competencies toward which your program is oriented. And, in the type of assessment we are proposing, student competency statements become, in effect, the basis for the needs assessment items. In fact, this part of the implementation process could just as well be called an inventory of student competencies–an inventory of where students are in competency development and where they would like to be in their competency acquisition.

To illustrate how the competencies you have chosen can be converted to needs assessment items, the following examples are provided. From those presented in the Appendix we selected one competency from grade 1, one from grade 8, and one from grade 11.

First grade

Competency:	Students will describe how exercise and nutrition affect their mental health.
Needs assessment:	I can tell how exercise and eating habits make a difference in how I think and act.

Eighth grade

Competency:	Students will analyze effective family relations, their importance, and how they are formed.
Needs assessment:	I can tell why good family relations are important and how they are formed.

Eleventh grade

Competency:	Students will evaluate the need for flexibility in their roles and in their choices.
Needs assessment:	I can explain the need for flexibility in my roles and in my choices.

The actual assessment can be done using a card-sort approach or a questionnaire format. The card-sort approach provides direct interaction with students, but it takes more time to administer and score. The questionnaire format is easier to administer and score, but it does not provide the direct contacts with students some may desire.

To show you what a needs assessment questionnaire using converted competencies looks like, we chose a section of one used for grades 10, 11, and 12. The section of the questionnaire we chose is on life-career-planning competencies. See Chart 4.1.

To fill out this questionnaire, students were asked to complete the following steps.

Step 1: *Read each sentence carefully.* Each one describes things students are able to do in order to demonstrate learning in that particular area.

Now make a decision. Are you able to do what the sentence describes? Are you not able to do what the sentence describes?

Fill in the circle that shows what you think.

If you think you are to do what the sentence describes . . .

fill in circle *a* ⓐ ⓑ | ⓒ

If you think you *are not* able to do what the sentence describes . . .

fill in circle *b* ⓐ ⓑ | ⓒ

Step 2: *Choose the five sentences on each page that you would really be interested in learning to do.* Some of the statements will really interest you and some will not. *In the second column (circles are lettered* c*), fill in the circle for each of the five statements that you feel you need to learn how to accomplish.*

For example, if you feel you would really be interested in learning to analyze how characteristics and abilities develop, you would mark as follows:

I CAN 1. describe and analyze how an individual's characteristics and abilities develop.

ⓐ ⓑ | ⓒ

CHART 4.1

Life Career Planning

The statements on this page are about things you can do to show you are preparing yourself for the future and are able to make decisions about what you want to do in your life. Review your instruction sheet carefully. Then fill in the circle that shows what you think about that statement.

I CAN			
1. evaluate the importance of having laws and contracts to protect producers.	ⓐ	ⓑ	ⓒ
2. provide examples of decisions I have made based on my attitudes and values.	ⓐ	ⓑ	ⓒ
3. analyze the decision-making process used by others.	ⓐ	ⓑ	ⓒ
4. distinguish between alternatives that involve varying degrees of risks.	ⓐ	ⓑ	ⓒ
5. evaluate the importance of setting realistic goals and working toward them.	ⓐ	ⓑ	ⓒ
6. describe my rights and responsibilities as a producer.	ⓐ	ⓑ	ⓒ
7. explain and analyze how values affect my decisions, actions, and life-styles.	ⓐ	ⓑ	ⓒ
8. identify decisions I have made and analyze how they will affect my future decisions.	ⓐ	ⓑ	ⓒ
9. analyze the consequences of decisions others make.	ⓐ	ⓑ	ⓒ
10. explain how my values, interests, and capabilities have changed and are changing.	ⓐ	ⓑ	ⓒ
11. speculate what my rights and obligations might be as a producer in the future.	ⓐ	ⓑ	ⓒ
12. summarize the importance of understanding my attitudes and values and how they affect my life.	ⓐ	ⓑ	ⓒ
13. use the decision-making process when making a decision.	ⓐ	ⓑ	ⓒ
14. provide examples and evaluate my present ability to generate alternatives, gather information, and assess the consequences in the decisions I make.	ⓐ	ⓑ	ⓒ
15. assess my ability to achieve past goals and integrate this knowledge for the future.	ⓐ	ⓑ	ⓒ

It is common practice to describe needs assessment as a way of determining the discrepancy between what exists and what is desired. If this practice is observed rigidly, only contemporary needs will be recognized, and the needs of the past, or those which have been filled, may be overlooked. When asked to respond to a need statement, individuals would be justified in asking whether it made a difference if the statement represented a need they felt was important but was being satisfied, or if it was a statement that represented an unmet need. From a program-planning perspective it is important to know which needs are being met as well as which needs deserve additional attention.

The opportunity to respond to a relevant sampling of needs is another important point. Simply stated, How can a need be identified if no one presents the statement? Limited coverage, insignificant choices, or redundancy may distort a needs survey. As a result it is often advantageous to use an adopt-adapt strategy for selecting and modifying needs statements from existing instruments rather than constructing new ones. This avoids the possibility of overlooking important need areas and insures a broad spectrum of possible needs to be assessed.

A final point to keep in mind deals with the question of who should be assessed. The answer is anyone who is involved in the educational process, including those responsible for receiving education. This includes students, educators, parents, community members, employers, and graduates.

Students: This group should receive top priority in any needs assessment. Who knows more about students than students? Students can tell you what they need as a group and as individuals. They will also let you know whether or not the current program is meeting their needs.

Educators: Assessing this group will give you their perceptions of student needs as well as perceptions of their own needs.

Parents: This group will help you identify what they feel their children should learn from school experiences. Including them in the needs-assessment process offers them an opportunity for involvement in planning the guidance program. As a result of their personal involvement they may be more willing to offer their support.

Community members: Included in this group are individuals who are not employers; yet, they support the school financially. Information from this group may give a somewhat different perspective to the same type of information gained from an assessment of parents.

Employers: Those who are responsible for hiring graduates of your school system or for hiring students still in school have definite ideas about the outcomes of education they expect. Including employers in the needs assessment process will give the school an opportunity to know what employers expect as well as offering the employers a chance to know more about the guidance program.

Graduates: An assessment of this group can provide information about the effectiveness of the guidance program for those who are applying their skills in post-high-school pursuits. They can help identify areas that are of the most benefit as well as areas that need strengthening.

Because of time and resource limitations, you may not be able to assess all of these groups. If you must restrict the number of groups to be assessed, students and educators should receive attention first since they are the ones most immediately involved. It may be that you could assess students and educators the first year and then assess members of the other groups in following years. It is important, however, that each of these groups be assessed at some point in the continuing needs-assessment process.

Decide What Stays and What Goes

By this time you have documented something you already knew: that a number of activities for which you are responsible are not guidance activities at all or are, at best, only tangentially related to guidance. They have become a part of the guidance program over the years, perhaps by someone's design but more likely by default. As you know, too, no matter how they became part of the program, once they are, it is difficult to remove them. And, what is worse, these responsibilities consume valuable time and resources, time and resources that are needed for conducting the actual guidance program.

In addition, you are now aware more than ever before, that modifying an existing program is more difficult than beginning a new program. The task of modifying a program is complicated for many reasons, but particularly because you cannot take a vacation from traditional duties while the modification process takes place. In a very real sense you are caught in the situation of remodeling your program while you are working in it.

Thus, we arrive at one of the most crucial steps in the entire program improvement process. The questions to be answered now are, What stays? What is modified? What goes? and, What happens to what goes? To help answer these questions, two sets of data collected previously are used: data from the assessment of the current program (see Chapter 2) and data from the assessment of perceived student needs. Both sets of data are used in the context of the newly selected structure for the guidance program. The basic strategy to be employed is called the *displacement strategy*, in which desired guidance program activities replace undesired or inappropriate activities or duties.

Support for the displacement strategy is provided from two types of analyses. The first is staff time analysis, and the second is discrepancy analysis. Both approaches are used to provide rationale for suggested program changes and to point out program directions for the future.

Staff Time Analysis

One basis for supporting the displacing of undesired or inappropriate duties with desired guidance activities is to examine the amount of staff time required to accomplish the undesired or inappropriate duties. This is one place where the two sets of data collected previously are used. This was done, as you will recall, by the guidance staff of the Mesa, Arizona, Public Schools as reported in Chapter 2 (Mesa Guidance and Counseling Department 1973). They analyzed where they were spending their time by keeping track of their activities and the time spent completing them. Then, to present their findings, they used a program structure with four outcome domains: academic learning, educational-vocational, interpersonal, and intrapersonal. In cataloguing their use of time, they used four domains as well as several other categories. Table 4.1 presents the distribution of counselors time at the high school level based on their assessment of the time needed to carry out their current activities.

By keeping a log, the guidance staff in the Mesa Public Schools found they were spending only 25 percent of their time carrying out activities related to the improved program they were proposing. Seventy-five percent of their time was devoted to administrative-clerical duties. The staff had previously decided that

TABLE 4.1

Distribution of Staff Time in the Current High School Program

Area	Percentage
Academic learning counseling —helping students to learn better in school and elsewhere —helping them improve their student skills and habits	5
Educational-vocational counseling —helping students plan better both their current and future schooling and work	10
Interpersonal counseling —assisting students to get along better with others	5
Intrapersonal counseling —facilitating students to feel better about themselves as individuals	5
Conducting student registration—schedule changing and orientation	66
Handling attendance concerns	5
Conducting and receiving in-service training	3
Supervising clubs	1
	100

Courtesy of Guidance Department, Mesa Public Schools, Mesa, Arizona. Byron E. McKinnon, Director.

their improved program would focus on helping students achieve outcomes in the four domains previously identified. The staff had also previously conducted an assessment of student needs, using items derived from the chosen goals and objectives of the improved program. Considering the results of this assessment, it was obvious that a different distribution of guidance staff time was required. Table 4.2 presents their recommended time distribution for the high school.

As you can see from a comparison of Tables 4.1 and 4.2, the staff decided their goal would be increase time spent on activities in each of the student outcome domains of the improved program. Previously they determined they had spent about 25 percent of their time doing that. For the improved program they felt they should be spending approximately 70 percent of their time helping

TABLE 4.2

Distribution of Staff Time in the Improved High School Program

Area	Percentage
Academic learning counseling	10
Educational-vocational counseling	20
Interpersonal counseling	20
Intrapersonal counseling	20
Conducting student registration	0
Handling attendance concerns	0
Conducting and receiving in-service training	5
Supervising clubs	0
Maintaining guidance program organization and implementation	10
Engaging in program research and development	15
	100

Courtesy of Guidance Department, Mesa Public Schools, Mesa, Ariaona. Byron E. McKinnon, Director.

students achieve outcomes in the four domains. The move toward 70 percent was supported by the data from the assessment of student needs. In addition, it was decided that scheduling, handling

TABLE 4.3

Distribution of Staff Time in the Current Junior High School Program

Area	Percentage
Academic learning counseling	12
Educational-vocational counseling	10
Interpersonal counseling	36
Intrapersonal counseling	27
Providing school support—such as conducting registration and handling noncounseling activities	10
Providing teacher-staff counseling	3
Conducting community activities	2
	100

Courtesy of Guidance Department, Mesa Public Schools, Mesa, Arizona. Byron E. McKinnon, Director.

attendance, and supervising clubs would be eliminated as activities in the improved guidance program. Two additional areas, continued program development and research, were added.

So that you can see the full scope of the program K–12, the same information that was just presented for the high school level

TABLE 4.4

Distribution of Staff Time in the Improved Junior High School Program

Area	Percentage
Academic learning counseling	13
Educational-vocational counseling	9
Interpersonal counseling	33
Intrapersonal counseling	33
Providing school support—such as conducting student registration and handling noncounseling activities	7
Providing teacher-staff counseling	3
Conducting community activities	2
	100

Courtesty of Guidance Department, Mesa Public Schools, Mesa, Arizona. Byron E. McKinnon, Director.

TABLE 4.5

Distribution of Staff Time in the Current Elementary School Program

Area	Percentage
Academic learning counseling	10
Educational-vocational counseling	5
Interpersonal counseling	50
Intrapersonal counseling	20
Conducting student registration	2
Handling attendance concerns	3
Conducting and receiving in-service training	10
Supervising clubs	0
	100

Courtesy of Guidance Department, Mesa Public Schools, Mesa, Arizona. Byron E. McKinnon, Director.

is also presented for the junior high school and elementary school levels. Tables 4.3 and 4.4 present the distribution of staff time for the current and recommended junior high program while Tables 4.5 and 4.6 present the distribution of staff time for the current and recommended elementary school program.

TABLE 4.6

Distribution of Staff Time in the Improved Elementary School Program

Area	Percentage
Academic learning counseling	5
Educational-vocational counseling	10
Interpersonal counseling	10
Intrapersonal counseling	10
Conducting student registration	0
Handling attendance concerns	0
Conducting and receiving in-service training	15
Supervising clubs	0
Maintaining guidance program organization and implementation	30
Engaging in program research and development	20
	100

Courtesy of Guidance Department, Mesa Public Schools, Mesa, Arizona. Byron E. McKinnon, Director.

Discrepancy Analysis

Another basis for supporting the displacing of undesired or inappropriate duties with desired guidance activities is the results of a discrepancy analysis. Is there a discrepancy between what you want your guidance program to accomplish and what your guidance program seems to be accomplishing? A discrepancy analysis will help you formulate statements that can be used to support program changes. Information gained from the perceived needs assessment can be contrasted with information obtained in the current program assessment (Chapter 2). This procedure allows you to compare your desired guidance outcomes and activities with the actual guidance outcomes and activities that currently exist. If this comparison results in a significant discrepancy, a statement reflecting this analysis can be used as authoritative fact. The following example illustrates this procedure:

Preferred student outcome: Students, parents, and educators agreed (perceived needs assessment) that prior to exiting the eleventh grade, all students should be able to demonstrate career decision-making skills within or exceeding their range of expectancy.

Current student outcome: May 1980 test results indicate that out of 600 eleventh-grade students in Boone County, 390, or 65 percent, are achieving within or exceeding their expectancy range in demonstrating career decision-making skills.

Detected discrepancy: 35 percent, or 210 eleventh-grade students.

Data source: Career Skills Assessment Program, Test 3—Career Decision-making Skills, College Entrance Examination Board, 1978.

Description of assessment population:

Type of student: All students, except those classified under special education categories.

Level of students: Grade 11.

Number of students: 600.

Special characteristics: Sex.

Based on this analysis, the following questions are asked and answers provided concerning the impact of decision-making activities for eleventh-graders.

Question:	To what extent does the data indicate that the detected discrepancy affected a large proportion of the assessment population?
Analysis technique:	Mathematical subtraction between preferred and current outcome statements.
Finding:	35 percent of the assessment population was affected.
Question:	To what extent does the data show that the detected discrepancy has persisted across time?
Analysis technique:	(a) Examination of student performance data for past three years.

	(b) Comparison of summarized historical data with desired student outcome statements.
Finding:	Detecting discrepancies of approximately same magnitude for the last three years.
Question:	To what extent does the data show that the detected discrepancy was substantial?
Analysis technique:	Distribution of assessment population by race, sex, expectancy range; reported in numbers and percentages.
Finding:	35 percent of the population was deficient, thereby evidencing a substantial discrepancy.
Question:	To what extent is the detected discrepancy selective? That is, was it found to be more highly associated with certain segments of the population than others?
Analysis technique:	Distribution of assessment population by sex and expectancy range; reported in numbers and percentages.
Finding:	Male population most affected.

These findings suggest that although a majority of eleventh-graders demonstrated career decision-making skills as measured by Test 3 of the Career Skills Assessment Program, a substantial number, 35 percent, were deficient in those skills. Moreover, among the 35 percent who were deficient, males were most affected. This discrepancy, along with those identified across the other activities of the guidance program, serve as beginning points for program modification (displacement) particularly when these data are combined with other data sources, such as staff time analysis and needs assessment.

To illustrate this point, suppose you used staff time analysis and discrepancy analysis to help you decide what goes and what stays in your program. From the discrepancy analysis you found out that 35 percent of your eleventh-graders were unable to demonstrate career decision-making skills within or exceeding their range of expectancy. An inspection of staff time revealed that the guidance personnel responsible for conducting decision-making activities in the eleventh-grade had spent an inordinate amount of time doing individual credit checks of eleventh-grade transcripts in preparation for planning twelfth-grade coursework. Add to this the information from the needs assessment that revealed that eleventh-graders, particularly males, felt they wanted more help on increasing their decision-making skills. Armed with these results, you are in a position to make an authoritative statement concerning what activities should be displaced. What should be eliminated in this situation, in our opinion, would be the individual eleventh-grade credit checks by guidance personnel. This would be replaced by guidance personnel spending more time assisting eleventh-graders, particularly males, to work on their decision-making skills. Credit checks could be shifted to become the responsibility of advisors in an advisory system.

Decide How to Handle Displaced Activities

The question of who handles displaced activities if the guidance staff does not is an issue that is not often discussed but should be. In the foregoing example, three types of data were combined to suggest that the time of guidance personnel might better be spent working directly with students to improve their decision-making skills rather than to do credit checks. It was further suggested that credit checks could and should be incorporated into the role of advisor in an advisory system. In this way, credit checks became an all-staff responsibility, not a time-consuming activity for one group of staff members.

Another activity that most guidance staffs would like to displace is scheduling, or student registration. Note that the guidance staff of the high school in Mesa proposed that they drop student registration from their list of activities (see Tables 4.1 and 4.2). This raises the following questions: If the guidance staff doesn't do scheduling, who will? and, How will this necessary activity be completed? One answer to these questions is to say that this activity is a total school staff responsibility and that the various departments in the school are responsible to work together to accomplish it. For example, the vast majority of scheduling or registration can be done in one day using a university-type scheduling system. It does not have to be done one student at a time by guidance staff over a period of three to four months. Many school districts have found this out already and have moved to an all-staff-involved student registration.

Another problem area for counselors is cumulative records. Often counselors are asked to maintain the cumulative records, to become in effect the school registrar. Those of you who have this responsibility know that it is time consuming. The answer to this problem is to work toward hiring a registrar or at least clerical personnel to do the job.

A more recent problem for many counselors, particularly at the elementary level, is P.L. 94-142, the Education for All Handicapped Children Act. In many states there is legislation on the same topic. The cornerstone of P.L. 94-142 is the *individualized education programs* (IEPs), which must be developed for individuals and reviewed annually. Often school counselors have been given major responsibility for the IEP development and all of the coordination that this requires. While counselors have a key role to play in this process, it is not their role entirely. A recent survey of ASCA members on the impact of P.L. 94-142 indicated that a significant number perceive the need for additional personnel and materials to comply with the law (Lombana and Clawson 1978).

As you are considering these problem areas that have just been discussed as well as those you may have in your own school district, let's go back and again consider the basis on which decisions are made to displace activities. The basis for these decisions begins with the assessment of your current program. You will recall this involved (1) identifying current guidance activities; (2) identifying student competencies that result from those activities; (3) cataloguing current resources; (4) gathering perceptions about the current program; and (5) keeping track of counselor time.

Then, using the information collected from the assessment of the current program and the information obtained from reviewing a program model, you selected the structure for your improved program. Next the final list of student competencies K–12 was agreed upon. This final list was the result of keeping those competencies from the current program that fit into the new program and adding others to round out the final list.

Once this was done, guidance activities and resources were matched to the competencies on the basis of how well they assisted students achieve these competencies. Activities from the current program that fit were kept. Some new activities and resources were probably added. Those activities from the current program that didn't fit became the target for displacement. The assessment of student needs (competencies) provides data from students, parents, staff, and the community about the importance of the competencies and the corresponding appropriate activities and resources. The assessment also provides data about which competencies (and hence activities and resources) should be emphasized at which times in the overall program. Staff time analysis and discrepancy analysis provide ways to analyze data and develop rationale for program changes.

Develop Timetables and Activity Schedules

Now that you have chosen a way to organize your improved guidance program, made decisions concerning student competencies to be achieved, and chosen appropriate guidance activities and resources to be incorporated into your program, the next step is to establish timetables and activity schedules. This is another one of those important steps in the program improvement process because it provides still another opportunity for the guidance staff to make their program visible—visible in the sense that others can see the totality of the program and the resources and time required to carry it out. This step is also important because of the discipline required to lay out timetables and activity schedules. It forces you to think through carefully how the program will unfold before it actually does.

To accomplish this task we recommend that timetables and activity schedules be developed for each of the major components of the program. In the case of the components suggested in this chapter, that would be the *guidance curriculum, individual planning, responsive services, and system support.* Since the guidance curriculum is the largest component in terms of competencies, activities, and resources, we suggest you start with it first. Another reason for starting with it first is that it contains the most activities and resources that will require the cooperation and assistance of other staff; hence longer-term planning is needed.

The first step, then, is to develop a timetable chart for the guidance curriculum by grade-level groupings (see Table 4.7).

List the months of the school year, or the entire year if summer work is planned, across the top of the chart and the activities to be accomplished through the curriculum down the side. Then time lines are laid out showing when an activity would begin and when it would end. When completed, each of the grade levels would have a guidance curriculum timetable.

TABLE 4.7

Guidance Curriculum Timetable—Grade 7

Major Activities	Aug	Sep	Oct	Nov	Dec	Jan	Feb	Mar	Apr	May	Jun	Jul
1.												
2.												
3.												

Based on this overall picture, more specific activity schedules (some planners call them *task-talent-time charts*) are developed. These charts are also labeled by the grade for which the activities are scheduled. They begin with the domain, goal, and competencies, identify the target group; and then specify the activities to be used, when they will be done, and by whom.

Continue Your Public Relations Activities

In Chapter 1 we suggested that you begin public relations activities at the same time you begin the overall guidance program improvement process. As you will recall, the beginning point was the development of a public relations plan that would unfold as an integral part of the program improvement process. One part of the plan probably included organizing and giving presentations to various school and community groups. Here are some ideas on planning and giving presentations:

Planning Presentations

There are many public relations strategies that can stimulate public awareness, understanding, and acceptance of the guidance program. However, remember that a *personal approach*, an appeal to a particular individual or group through a direct personal presentation, is more likely to get a favorable response than an impersonal appeal. Also, remember that effective communication means tailor-made presentations especially designed for the situation, time, place, and audience.

When you plan a presentation, you should give careful attention to the art of communication. The following list of communication elements are suggestive of points that should be kept in mind in planning presentations.

The 7 C's of Communication

1. Communication starts with a climate of belief. This climate is built by performance on the part of the practitioner. The performance reflects an earnest desire to serve the receiver. The receiver must have confidence in the sender. the receiver must have a high regard for the source's competence on the subject.
2. A communications program must square with the realities of its environment. Mechanical media are only supplementary to the word and deed that takes place in daily living. The context must provide for participation and playback. The context must confirm, not contradict, the message.
3. The message must have meaning for the receiver, and it must be compatible with the receiver's value system. It must have relevance. In general, people select those items of information which promise them greatest rewards. The content determines the audience.
4. The message must be put in simple terms. Words must mean the same thing to the receiver as they do the the sender . . . the farther the message has to travel, the simpler it must be. An institution must speak with one voice, not many voices.
5. Communication is an unending process. It requires repetition to achieve penetration. Repetition—with variation—contributes to both factual and attitude learning. The story must be consistent.
6. Established channels of communication should be used—channels that the receiver uses and respects. Creating new ones is difficult. Different channels have different effects and serve effectively in different stages of the diffusion process.
7. Communication must take into account the capability of the audience. Communications are most effective when they require the

> least effort on the part of the recipient. This includes factors of availability, habit, reading ability, and receiver's knowledge.[1]

Once you have the 7 C's of Communication in mind, the next step is to lay out your presentation systematically in outline form. The first step is to consider who your audience will be. This will determine to a large extent what your content will be, how you will organize the presentation, and the type of media to be used. For example, if you are making a presentation to teachers in your school, you can probably be more technical than if you are speaking to a local service organization.

Although the content of your presentation is determined in part by your audience, there is a core of material from which all presentations will be prepared. It is important that the core material reflects your improved program—a comprehensive guidance program based on needs assessment data, organized around individual outcomes, containing appropriate evaluation and feedback procedures.

The next step in planning is to consider how you will organize your presentation and what type of presentation you will use. A wide range of presentation methods are open to you, so don't limit yourself only to the traditional lecture. Think about the variety of presentation methods, such as forums, panels, symposia, discussions, or group interviews, and then think about ways of effectively combining them for your presentation. As you do, also think about the sequence of events (content) within the planned presentation so you can relate a particular aspect of presentation content to the most appropriate presentation method. Some kinds of content lend themselves to a straightforward presentation, perhaps by one person, while other kinds of content might best be conveyed through panel discussions or testimonials of students or staff.

A final step in planning is to consider the media you will use. The impact that media has on audiences and content is

[1] Scott M. Cutlip, Allen H. Center, *Effective Public Relations*, 4th ed., ©1971, pp. 260-61. Reprinted by permission of Prentice-Hall, Inc.

well understood. As a result careful thought must be given to the kind of medium best suited for your presentation content and your audience. Again, there is a wide range of media available to you, ranging from motion pictures to handouts. Ordinarily your time will be limited, so don't think in elaborate terms. Rather, stay with simple transparencies, charts, and brief handouts. Whatever medium you use, have it prepared in sufficient time to try it out on several groups similar to your intended audience. Medium is an excellent assistance to your program if it works; if it doesn't work, it will be a detractor.

To summarize, here are some key points to remember in planning a presentation:

---

1. Carefully delineate the presentation method to be used and the steps which need to be taken to use those strategies effectively.
2. Check to see if you are covering all the principal components of your program.
3. Develop a complete outline of all the organizational aspects of the presentation.
4. Develop plans for the development and use of media.

Giving Presentations

Effective planning is the foundation of a good presentation. There are a number of points to keep in mind, however, just prior to, during, and after a presentation.

Just prior to a presentation use these questions as a way of checking your presentation readiness.

---

1. Is the room arranged and lighted properly?
2. Is the room ventilation adequate?
3. If name tags or signs are to be used, are they available and in place?
4. Are the audio-visual resources functioning properly?
5. Is the media (handouts, transparencies) ready?

6. Will the presenters be in place and ready to begin on time?
7. Is someone available to handle such duties as turning out the lights, getting more chairs, and closing doors?

During a presentation, keep these points in mind:

---

1. Look at your audience.
2. Avoid reading your presentation word for word.
3. Be enthusiastic.
4. Use common language; seek audience identification in vocabulary and anecdote.
5. Stick to time limits.
6. If you are a panel member, give your attention to the person speaking; it helps to smile occasionally too!
7. Speak clearly and loudly enough to be heard by all.

In summary, audiences respond favorably to presentations in which presenters use straightforward English and simple phrases, know their subject, stay within the announced time limit, and are enthusiastic. If possible, the audience should be allowed sufficient time to ask questions. Give direct, straightforward answers. If questions are asked about sensitive topics, offer to meet with the questioner after the meeting.

Finally, use questions such as these to evaluate your performance: Was the presentation adequately planned? Did the audience understand the message? What could have been done differently? Was the audience reached? Could better provisions have been made to handle unforeseen circumstances? Were the provisions made in advance for measuring presentation impact? Were such provisions adequate?

Initiate Staff Development Activities

In the subsection of this chapter on developing timetables and activity schedules (pages 132–133), it was recommended that timetables and activity schedules be developed for all of the program components of your improved program. If you used the components recommended in Chapter 3 to structure your improved program, then you will have a system-support category and, within it, staff development activities. You also will have developed a timetable and activity schedule for these activities. If you haven't done so, we recommend that you do so now.

As you are planning and carrying out staff development activities, there are a number of points you may wish to consider. These include assessing staff competencies, identifying sources of staff development expertise, considering delivery methods, selecting appropriate times for staff development, and staff development evaluation.

Assessing Staff Competencies

One reason some program improvement efforts fail is because sufficient attention is not given to staff needs. Sometimes staff are asked to conduct activities for which they have had no training, or the training they had occurred early in their careers, before some of the newer techniques and resources had been developed. Thus it is important to assess staff competencies to find out the competencies they have as well as the competencies they would like to work on. To assess staff competencies the same methodology can be used that was used to assess the perceived needs of students; only the items are changed. The items to be used can be generated from the competencies that have been identified as necessary to organize and implement an improved guidance program. Then a card-sort or questionnaire is developed. The results can be tabulated using the same procedure used for the assessment of perceived student needs.

Identify Sources of Staff Development Expertise

A major task in staff development is to identify available staff development expertise. There are many possible sources of expertise for staff development including current staff members, community members, colleges and universities, professionals, professional organizations, commercial firms and outside consultants, intermediate school district offices, the state Department of Education, other school districts, and business and industry personnel.

Once you are aware of the sources of expertise in your area, your task is to decide which source is best for you. Sometimes combinations of sources may be used. Since cost may be a major factor in determining the source you may use, don't overlook the use of your current staff members or staff members from nearby school districts. Supervisory personnel in state departments of education, intermediate school districts, or larger school districts are also usually cost-free.

Choose Delivery Methods for Staff Development

The next step is to decide on the delivery method or methods to use. The decision on the type of method to use will be affected in part by the source of staff development expertise chosen. Just as there are many sources of expertise available, there are also a variety of delivery methods available, including lecturers; reading materials; graphics, auditory aids, films, slides, and filmstrips; demonstrations; programmed learning; discussions, simulations; and direct experience. The table that follows (Table 4.8) lists these eleven delivery methods and outlines the group size for which they are best suited, the cost facilities and equipment needed, major features, and some cautions to observe when using them.

Selecting Appropriate Times for Staff Development

Careful consideration should be given to when staff development activities take place. Some writers suggest, with good reason, that late afternoon or evening sessions are to be avoided if possible. During-school sessions are recommended. This is something, however, you will have to determine based on your staff and their life-styles.

We recommend that staff development activities be interspersed over the length of the program improvement process. Look at periods of time that tend to be down times—February, for example—and plan special events to build morale. Use such incentives as district in-service credit, college or university credit, and existing district in-service days. Make sure that staff development activities are planned well in advance so that all involved can participate in them.

TABLE 4.8

Delivery Methods

Method	Group Size	Cost	Facilities and Equipment	Major Features	Caution
Lecturers	• Should be considered when group is over 20 people	• Depends on speaker (free to $1,000 per day) • Cost may be shared with other districts	• Adequate size room • Microphone • Visual aids • Comfortable seats	• Useful in introductions or overviews • Effective when providing factual information or explanations • Use when gap between learner and lecturer is large • A lecturer can cover in 5 minutes what it might take 30 minutes to cover in a group	• Boring if speaker not organized or not easily heard • Provisions need to be made for feedback and practice • Lecturers impart knowledge more easily than they change attitudes
Reading materials	• Individual	• Cost varies • May be bought or borrowed	• Access to journals, ERIC, conference reports, monographs, dissertations	• May provide an overview and specific information • Best suited for a self-motivated learner	• Materials may be difficult to obtain • Provisions need to be made for feedback and practice
Graphics (including charts,	• Can be adjusted to	• Can be elaborate or done simply	• Made from whatever materials avail-	• Indicate basic relationships	• Should be brightly colored to arouse interest

graphs, bulletin boards, posters, and cartoons)	large group, small group, or individual needs	and locally for less	able • Need display area	• Help staff recall key concepts • Arouse interest • Serve as productive room decorations	• Should be large and simple enough to be understood by all
Auditory aids (including cassettes, tape recorders, and record players)	• May be adjusted to large group, small group, or individual needs	• Commercially produced tapes and records more expensive than locally developed tapes • Cassette players, record players, and tape recorders may be rented	• Players for tapes or records • Storage for tapes or records • Storage for tapes, records, and players	• May provoke interest by bringing a slice of life into the classroom • May be stopped and played again	• Check for relevance and clarity of sound reproduction • Maintain machines to prevent mechanical failure • Make provisions for participants to react to presentation
Films, slides, and filmstrips	• Can be adjusted to large group, small group, or individual needs	• Commercially produced film averages $1,000 per running minute • You may rent films for less • You may share costs with other districts	• Production and storage facilities • Projector • Screen • Dark room • Adequate acoustics	• Useful as an introduction or overview • May overcome language and experience barriers • Provide views of action difficult to observe first hand • May be stopped, and played again • Often used to motivate, to dramatize, or to pose a problem	• Check commercially made for relevance • Maintain machines to prevent mechanical failure • Make sure all participants can see and hear • Make provisions for participant to react to presentation

TABLE 4.8 (Continued)

Delivery Methods

Method	Group Size	Cost	Facilities and Equipment	Major Features	Caution
Demonstrations (including exhibits and field trips)	• Can handle large numbers of people—often in groups	Vary with • travel • lodging • entrance fees • consultant fees	• Room where all can see and hear • A camera or recorder to document demonstration	• Motivate staff • Expose staff to new methods and materials • Prepare staff to try out new skills	• Check demonstration content for relevance before attending and again before applying what you saw or heard to your particular situation
Programmed learning (students works his way through a series of small steps with built-in feedback)	• Individual	• Cost varies • Expensive to develop unless used by large numbers of people	• Books • Learning machines	• Suited to a self-motivated learner who has needs in a specific field that are not necessarily related to the needs of other staff members	• Check for relevance
Videotapes	• May be adjusted to	• Expensive, but may be shared	• Camera • Tapes	• Staff members' performance may be played	• Staff must agree to being videotaped

	large or small group needs	with other districts or rented	• Playback equipment • Storage facilities	back and analyzed • Tape may serve as a model for new staff members	• Equipment must be maintained to guard against mechanical failure
Discussions (including conference, buzz sessions, and brainstorming)	• 2–20 people in any one group	Depends on • leader's fees • elaborateness of props	• Quiet, comfortable location • Centralized conferences may have better facilities and equipment than local districts can afford	• Share, develop, and refine participants' attitudes and skills	• Can get off the subject and be time consuming if not well directed • Topic should be investigated for relevance before attendance and again before generalizations are made that affect your program • Group members must learn to tolerate differences in opinion
Simulations (including laboratory experiences, role playing, case analyses, in-basket techniques)	• 8–25 per group • Less than 8 people may provide too little group input	Depends on • leader's fees • elaborateness of props	• May be spontaneous or structured with script and props	• Conceptualize what participants already know • Work on attitudes as well as skills • Participants should have some basic knowledge of the skills involved	• Need to create an open atmosphere in which participants are comfortable sharing their attitudes and skills

TABLE 4.8 (Continued)

Delivery Methods

Method	Group Size	Cost	Facilities and Equipment	Major Features	Caution
Direct experience (including apprenticeships, job rotations, and supervised implementation	• Individual	• Pay as you go and/or • You pay a university to supervise you	• A program similar in resources, students, and objectives to the one in which you will serve	• In-depth learning • High degree of retention • May be for varying lengths of time	• Contract with employer or supervisors to receive varied, relevant training in specific areas • Make sure program in which staff member is training is using valid, up-to-date methods • Provide for frequent checks on participant's progress

A Planning Model for Developing a Career Guidance Curriculum, Monograph 12, (Fullerton, California: California Personnel and Guidance Association, 1978), 52–53.

Evaluate Staff Development Activities

There are a number of ways to evaluate the impact of staff development activities. These include the use of questionnaires, achievement-type tests, observation, and demonstrations. Whatever approach is used, it is necessary to match it with the actual staff development experience. For example, some evaluation devices are more appropriate for direct-experience staff development experiences than they might be for lectures.

MAINTAINING YOUR IMPROVED PROGRAM

As you know by now, organizing and implementing an improved guidance program is a difficult task. As difficult as it is, however, it is an even more difficult task to maintain an improved program; to maintain it so that it does not revert back to its original form. To assure that the program does not revert back, a number of activities require attention. These include providing reinforcement to staff, making program adjustments, continuing public relations activities, and continuing a staff development program.

Provide Reinforcement

Part of the difficulty of maintaining a program stems from problems associated with trying to change staff work behavior patterns. It is relatively easy to do something new once, particularly if it is highly visible. Also, some staff members may sabotage the improvement process by going along with a new activity once, but then, when it is over, they withdraw their support from repeating that activity. As a result, it is important to build into the program maintenance process, ways of assessing the need for reinforcement and ways to provide such reinforcement.

Since the need for reinforcement occurs over time, it is important that at least part of your staff development program be designed to provide for such reinforcement. Skill building and discussion sessions that take place on a regular basis are mandatory.

Such sessions should be planned to coincide with specific events in the guidance program. On a more informal basis, birthday parties, potluck dinners, and other socials are helpful ways to provide reinforcement. When you plan such events, consider the down times that occur in any academic year. For example, in some areas of the country, the last part of January and the first part of February need some special events to brighten the day.

Make Program Adjustments

As your program unfolds, there will be times when it is necessary to make program adjustments. Keep in mind that such adjustments are fine-tuning adjustments. They are not the major types made as a part of your initial program improvement. Also, any changes in your program should be made only after careful thought. Some needed changes will be obvious; others will not be obvious. As a rule, count to ten before making any substantial changes. Some activities need time to take hold and as a result may not show up too well at first.

The kinds of changes you can expect to be making most often as you are fine-tuning your program may include modification of timetables, modification of activity schedules, substitution or modification of activities, substitution of resources, changes in student competencies at various grade levels, modification of public relations activities, and changes in staff development programs.

Continue Your Public Relations and Staff Development Activities

To maintain your improved program, a continuation of your public relations activities is necessary. By this we don't mean to suggest that public relations alone will make a program successful. We do mean, however, that public relations are an integral part of a comprehensive guidance program. This point is made here to underline the importance of continuous public relations activities to the success of the guidance program. It is our way of saying keep up with what you have been doing. Public relations activities are needed over the lifetime of the program.

The same point that was made concerning your public relations program is made with the staff development program. Keep it going over the lifetime of the program. Modifications in the program will take place as new staff needs emerge. Staff development activities are so important because without them it is difficult for a staff to grow personally or professionally. That is why we are providing you with this reminder at this time.

REFERENCES

Cutlip, S.M., and A. H. Center, *Effective Public Relations* (4th ed.). Englewood Cliffs, N.J.: Prentice-Hall, Inc., 1971.

Gelatt, H.B., *Selecting Alternative Program Strategies. Module* 7, Palo Alto, Calif.: American Institutes for Research, 1975.

Lombana, J., and T. Clawson, "Counselors Surveyed on P.L. 94-142," *American Personnel and Guidance Association Guidepost*, August 10, 1978, p. 12.

Mesa Guidance and Counseling Department, *Toward Accountability*. Mesa, Ariz.: Mesa Public Schools, 1973.

Chapter 5

Conducting Evaluation

Since this is the last chapter of the book and it is on evaluation, evaluation is automatically the last step in the program improvement process. Right? Wrong! This statement is wrong because the entire program improvement process is evaluation-based. Evaluation is ongoing, providing continuous feedback during all steps of the process. Evaluation is not something done only at the end of a program in order to see how it came out.

The purpose of evaluation is to provide data to make decisions about the structure and impact of the program. The assessment of your current program described in Chapter 2 is an example of gathering data to make decisions about ways to improve your current program. It also provided you with a baseline against which you can make judgments about the impact of the program on students, staff, and the community.

TWO TYPES OF EVALUATION

For the purposes of this book, we describe two types of evaluation. The first type is called *process evaluation.* The second type is called *product evaluation.*

* Parts of this Chapter were adapted with permission from F. E. Wellman and E. J. Moore, *Pupil personnel services: A handbook for development and evaluation.* (Washington, D.C.: U.S. Office of Education, August, 1975).

Process Evaluation

Process evaluation is exactly what it says. It is evaluation of the processes used in a comprehensive, developmental guidance program to help students attain the outcomes (competencies) of the program (Upton, Lowrey, Mitchell, Varenhorst, & Benvenuti 1978). Process evaluation involves the collection of data that shows how the program was actually implemented; it pinpoints what was done to whom under what circumstances. As a result, process evaluation is sometimes called formative, or implementation evaluation.

The major mechanism used in process evaluation is a program monitoring and reporting system. You should establish such a system at the beginning of the program improvement process. The results will help you determine where more support is needed, where minor adjustments should be made, and whether or not nonproductive elements should be phased out. A word of caution. As you are getting this system under way, you will also be beginning preliminary work on product evaluation–most likely the gathering of pretest student outcome data. Don't allow this to obscure the importance of process evaluation data or to interfere with the procedures for collecting process evaluation data. One way to reduce interference is to assign process evaluation to one task group and product evaluation to another.

As the process evaluation unfolds, be alert to possible unanticipated side effects. Sometimes activities will create effects unforeseen when initially they were put into operation. The program monitoring and reporting system should be sensitive enough to pick up these effects so that they can be handled immediately or can be explained when they appear in later process or product evaluation results.

The following outline may serve as a checklist you may wish to use to collect data for your program monitoring and reporting system:

Program Monitoring and Reporting System
Brief Interview-Short Answer Report

Teacher or Counselor ______________________ Grade __________

Implementation dates ____________ Goal or competency ____________

1. Activity or process—What did you do?
2. Student involvement—What did they do?
3. Student reactions—How did they react?
4. Successes—What worked?
5. Greatest difficulty—What changes or deletions?
6. Unanticipated effects—What were they?
7. Evaluation—Did it achieve what you wanted?
8. Resources—What was used? How did they work?

Go ____________ No Go ____________ Change ____

We recommend that this checklist or one like it be used by the task group charged with the responsibility for process evaluation. We further recommend that a master time schedule be drawn up showing when during the year program monitoring will take place, where it will take place (grade level, classroom), and with whom. It is not necessary to monitor each guidance activity in the program at all grade levels with all students. A sample of activities across guidance program components, grade levels, and students is sufficient.

Product Evaluation

Student outcomes, or competencies, as we have chosen to call them, are the products of a comprehensive, developmental guidance program. "Product evaluation is the measurement of these outcomes both at strategic points during the implementation of the program and at completion of the program" (Upton, Lowrey, Mitchell, Varenhorst, and Benvenuti 1978, p. 57). Just as process evaluation begins at the beginning of the program improvement process, so too does product evaluation. Also, just as you developed a plan for process evaluation, so too do you develop a plan for conducting product evaluation.

As you are beginning the process of laying out a product evaluation plan, there are a number of decisions you should make. Decisions need to be made about how you will design the evaluation. You also will need to consider instrumentation, when you will collect data, who will collect data, and how these data will be analyzed.

Evaluation Design

Evaluation Based upon Predetermined Criterion Standard Comparisons. The process of specifying posttest performance expectations for students is one way to evaluate competency-based programs. This means that you need to establish minimally acceptable performance standards for a competency by indicating the percentage of students in the target population who must attain a particular outcome in order for the program to be considered successful. For example, if it is expected that 95 percent of the students will be able to select a course of study consistent with their measured interest and ability, then the minimum acceptable performance level has been established at 95 percent.

The specification of the minimally acceptable level should occur at the same time that the competency is stated initially. There are no hard-and-fast rules for deriving performance standards. Rather, they are usually derived from professional judgment based on the experience of staff members. Performance will vary across competencies, rather than be uniform. Factors to consider in setting the minimal performance level for an outcome include the judged importance of the competency, the place of the competency in the developmental sequence, and the probability of attaining the competency.

The next step in the evaluation of competency attainment consists of checking students posttest performance to determine whether the stated acceptable percentage of students did, in fact, attain each competency. Summary data for making this determination consist of a tally of the number of students attaining the criterion level and the computation of the percentage of the target group achieving the competency. When the sample is small, you can complete this process manually with a check mark in a "yes" or "no" column for each student to indicate attainment or nonattainment of the competency. With larger samples or where more detailed information is desired, you may wish to use a distribution with means and standard deviations; percentiles can be used as summary data.

Evaluation Based upon Pretest or Posttest Comparisons. Another method frequently used with a program is pretest-posttest comparison. Before-and-after data are collected prior to exposure to a guidance activity and upon completion of the activity; these are then compared. The observed differences in the two measures are then interpreted in terms of (1) the statistical significance of the change, (2) the percentage of students attaining a predetermined change standard, or (3) the comparison of change among program and control groups.

Evaluation Based upon Participant versus Nonparticipant (Control) Comparisons. The criterion referenced and pretest-posttest comparisons just discussed provide information that is particularly relevant for program development. The crucial questions regarding the cause of the observed performance or change, however, cannot be answered by these types of comparisons. The cause-effect questions are critical in program continuation or elimination decisions and necessitate comparisons of the performance or gains of participating subjects with that of nonparticipating subjects. These types of comparisons not only provide you with evidence of competency attainment of participating students but also support conclusions that guidance activities were the primary causitive factor in the observed outcomes (where significant group differences were observed).

Evaluation Based upon Responsive Observations. Product evaluation provides information to determine whether specified student competencies have been attained. Evaluation data that provide information about what was not predicted or what was unanticipated are also important. Data on unexpected side effects document unintended effects of process operations and process dynamics. The unanticipated outcomes may be either positive or negative. For instance, the predicted outcomes may be achieved but at an unusually high expense of students' time or control over their school day. On the other hand, some of the most valued outcomes of a program may not have been stated in the program planning. This type of evaluation not only looks at unanticipated results but also focuses on student and staff responses to their experiences in the program. Attitude surveys, structured reaction sheets, and case-study techniques can be used to collect this type of data.

Another reason for using a responsive type of evaluation is to provide for case studies that portray effects and impact in a natural and direct manner. Case studies can provide a feel for what has happened that cannot be transmitted by examining hard outcome data. Reports of unusual impact on individual students as reported in case illustrations can also be used effectively in communicating more generalized group findings to the public. Take care however, that this type of evaluation is not interpreted as the collecting of testimonies.

Selecting or Developing Instruments

We recommend the following guidelines in the selection or development of instruments for the collection of product evaluation data.

---

1. The expected outcome for each student competency should be measured as directly as possible.
2. The instruments for collecting evaluation data should be appropriate for the intended respondent in terms of content, understandability, opportunity to respond, and mechanical simplicity.
3. Directions for the administration, scoring, and reporting for all instruments should be clear, concise, and complete in order to insure uniformity and accuracy in data collection.
4. The time required for the administration, scoring, and reporting of evaluation instruments should be kept at the minimum in order to obtain reliable information.
5. Evaluation instruments should meet the tests of validity for the objective, reliability in producing consistent results, and feasibility for the operational situation.

Scheduling Data Collection

The data collection schedule for product evaluation should be set up prior to the initial date of the evaluation period and should specify (1) the objective for which data are to be collected; (2) the instrument(s) to be used; (3) the group(s) or individuals from whom data will be collected; (4) the time when data will be collected (pretest, posttest, end of year, and so forth) in relation to the process schedule; and (5) the person(s) to be responsible for the collection of the data. The evaluation design, including the types of comparisons to be made, will dictate most of the decisions relevant to the data collection schedule.

Evaluation data collected for groups to make pretest-posttest comparisons or experimental control-group comparisons need to conform closely to a time schedule related to the process period. Pretest, or baseline data need to be collected prior to the initiation of the activities, and posttest data need to be collected at a specified time after the completion of the activity being evaluated. Some designs may also require the collection of data at specified periods during the activity period or as follow-up some time after the completion of the activity. All such data need to be collected on a predetermined schedule so that all persons involved in the evaluation can make plans and carry out the data collection in accordance with the design.

Staffing for Data Collection and Processing

Adequate staffing to handle the evaluation, including data collection and data processing, is essential. Organize your work groups to (1) plan and coordinate data collection and processing; (2) conduct in-service training of teachers or others who will be responsible for the actual data collection; (3) administer the information-collecting instruments; (4) handle the clerical details of preparing and distributing instruments, collecting and organizing completed instruments, scoring and coding data for processing, punching data cards, preparing tables, and preparing evaluation reports; and (5) write and interpret evaluation reports.

The absence of adequate staffing for an evaluation is frequently the underlying cause for the breakdown of the whole evaluation process. Symptoms of inadequate staffing may appear in the form of (1) unmet schedules, (2) resistance from teachers, (3) errors in data processing, and (4) incomplete reports that are not communicated adequately to program and administrative personnel.

The staffing needed for the evaluation cannot be standardized due to the differences in the nature and the size of evaluation projects from school to school. Staffing to conduct a full-scale program evaluation for all grades in a large school system will obviously require more leadership and more person-hours than will the evaluation of one specific activity in one grade. In any case, large or small, comprehensive or specific, the success of evaluation is dependent upon the assignment of specific time to staff for the planning, implementation, and interpretation of the evaluation.

COLLECTING PRODUCT AND PROCESS DATA

All data must be collected in accordance with a data collection schedule and with proper administration of evaluation instruments. This process requires careful planning and full cooperation of all persons responsible for collecting or providing the needed information. The following suggestions may be helpful in implementing efficient and accurate data collection:

---

1. The purposes and details of the evaluation plan should be communicated to all staff members who will be involved in the evaluation process. The threat of evaluation and the added burden of another task can be eased by a full explanation and discussion of all details for implementation before assignments are received by teachers, counselors, and others. Workshops can be used to discuss the data collection schedule and the instruments that are to be used. A good technique to acquaint teachers and counselors with the evaluation instruments is to let them complete all of the instruments they will administer. Emphasis should be given to instructions for the administration of all instruments and the necessity for uniform adminstration for all respondents. Also, where observers are to be used, it is important that they have had thorough training in making and recording their observations.
2. All instruments and evaluation instructions should be prepared and assembled well in advance of the date for implementing the data collection. Careful planning of the logistics of collecting and processing evaluation data will help in avoiding delays and in assuring compliance with the data collection schedule.
3. All respondents (students, teachers, parents) should be informed of the purposes for collecting information, and confidentiality should be assured where appropriate. Steps should be taken to motivate students to the task of completing tests or other instruments, as would be done in any other school testing situation. The assumption is made that the responses to evaluation instruments represent the respondent's best effort and an honest response. Any steps, within defined limits, that can be taken to assure the validity of this assumption will increase the reliability and validity of the data collected.
4. The data collected should be identified properly with respect to target groups, date, and person responsible for collection of the data.

This simple precaution will help prevent lost and mislabeled data and will enable follow-up in case questions arise regarding the data.

5. Evaluation tests and other instruments should be scored and coded for processing as soon as possible after the data are collected. The prearranged coding plan should be followed and then rechecked to assure accuracy. Many instruments can be scored and cards punched by machine where the appropriate answer sheets have been used and the equipment is available. Planning for the use of machine-scored answer sheets and the related machine-punching of data will result in greater speed and accuracy in processing evaluation data for analysis. (School systems that do not have their own test-scoring equipment and personnel with expertise in data processing should seek assistance from colleges and universities or commercial agencies that serve their region.)
6. The collection of process evaluation data usually becomes the direct responsibility of the guidance staff. It is crucial that process descriptions be compiled at the time of, or immediately following, the activity. Where the process being evaluated extends over a period of weeks or months, the maintenance of activity logs often facilitates the final preparation of the process report. It is next to impossible to reconstruct the process description after the fact when detailed records were not maintained during the process period. The involvement of the guidance staff in the preparation of process instruments often reduces resistance to reporting process and at the same time acquaints the staff with the content and procedures of the process data collection. Also attention should be given to identifying target groups, specific activities used, and materials and personnel resources needed for the process. These details are absolutely necessary for an adequate process report.

ANALYZING PROCESS AND PRODUCT EVALUATION DATA

The evaluation design is the blueprint for the analysis of evaluation data. The analyses should follow the design in all details; however, additional analyses may be made where the data warrant and where observations of your staff suggest the need for analyses not included in the original design. For example, the original design may have specified the analysis of gains in occupational knowledge between experimental and control subjects, but staff may have observed that the reading ability of students seem to be related to the criterion outcome. In this case, additional analyses may be desirable in order to determine the extent to which the observed outcomes were attributable to level of reading ability and how reading ability might be taken into consideration in program planning.

The mechanics of completing the analysis of evaluation data are important to assure speedy and accurate feedback from the evaluation. Computer-assisted analyses are most desirable where a mass of data is involved. Computers, however, depend upon the use of a program appropriate to the analyses and with a system of checks for errors or inconsistencies in the raw data input. The services of a computer programmer who understands the data and the desired output is essential.

Some types of evaluation information are not easily adaptable to computer analyses and may in fact be more meaningful when analyzed by you and your staff. For example, subjective counselor reports of guidance activities or certain types of student behaviors may lose meaning if quantified for computer analysis. These subjective analyses may be critical in the interpretation of other outcome data. In addition, small samples of activities or students may not warrant the use of computer analyses and thus will need to be handled manually. In such cases precautions should be taken to reduce human error to a minimum by establishing checks and rechecks.

REPORTING PROCESS AND PRODUCT EVALUATION RESULTS

The reports of evaluation results should be addressed to those persons who have an interest in the basic evaluation questions asked in the evaluation plan. Such persons include district research personnel, program directors, teachers, counselors, the lay public, and funding agencies. The variance in the interests and level of research understandings of these audiences dictates that preparation of separate reports that are appropriate for each group. These diverse interests can be satisfied by preparing (1) a technical report that constitutes a full research report of the design, all statistical data, and the evaluative conclusions; and (2) a professional report that focuses on the conclusions regarding the effectiveness of program activities and recommendations for program emphases and modifications. The basic content of these reports is discussed briefly in the following sections.

Technical Reports

A technical report should be a complete description of the program being evaluated, the design of the evaluation, the results, and conclusions and recommendations. The following outline can serve for the content of the technical report of a comprehensive guidance program evaluation:

Program Description

This part of the report should describe the program being evaluated in sufficient detail for the reader to replicate the process as evaluated. The target groups, the specific guidance activities, and the personnel and facilities used should be described in detail.

Evaluation Design

The description of the evaluation design should include a description of procedures used to formulate the evaluation questions and the program objectives. The specific evaluation hypotheses, the comparisons made, the operational definitions or instrumentation, and the types of analyses made for each objective should be described in detail. The case for the design as an adequate approach to answering the evaluation questions should be established in this part of the report.

Evaluation Results

The results of the evaluation should be reported in complete detail in this section. Each goal, objective, or competency evaluated should be presented with the evidence that it was, or was not, achieved. A summary of relevant descriptive statistics, and of the statistical analyses to test outcome hypotheses, should be reported in proper table form. When a large number of statistical tables are needed to report the results, it may be desirable to place some of these tables in an appendix of the report.

Conclusions, Discussion, and Recommendations

This section of the technical report presents the evaluative conclusions regarding the achievement of the stated goals, objectives, or competencies. The discussion of the outcome findings and conclusions can include subjective explanations and additional hypotheses suggested by the evaluative data. Recommendations that are supported by the evaluation, and relevant to administrative and program decisions, make up one of the most important parts of the evaluation report. This section should provide answers to the basic evaluation questions and discuss the program implications of the findings. The strengths and weaknesses of the program should be identified as indicated by the results. Recommendations for program modifications, and the nature of such modifications, should be presented along with the justifications based on the observed outcomes. Also, this section may include a discussion of the relationship between cost and outcome. Were the results worth the cost?

Appendices to the Technical Report

Materials that illustrate, describe, and support the other sections of the technical report may be included as information for the reader. Forms and unpublished instruments should be included as a matter of record and for readers who may not be acquainted with the details of the methods used. Also, detailed descriptions of activities may be included in an appendix if presentation in the body of the report would distract from clarity and readability.

Professional Reports

Reports of the evaluation for the professional and administrative staff of the school should be short and concise. Those interested in the details that support this report should be referred to the technical report. The professional report should include a brief summary of the findings, conclusions, and recommendations. Often most of this report can be taken directly from the conclusions, discussion, and recommendations section of the technical report. Statistical tables should be used only if absolutely necessary to document the outcomes summarized. However, summary charts that symbolically or graphically show the outcomes may be quite helpful. Technical language and reference to specific instruments should be avoided whenever possible. For example, it would be better to say the students had increased career awareness than to say the posttest scores on the vocational knowledge inventory were significantly higher than the pretest scores. This report should communicate in straightforward language what happened to students who received specific guidance activities. Sometimes uncluttered graphs or charts can be used effectively.

USING PROCESS AND PRODUCT EVALUATION REPORTS

Evaluation reports can be used for a variety of purposes. These include (1) conducting staff development, (2) making program decisions, and (3) making administrative decisions. Each of these uses is discussed in the following sections.

Staff Development

Evaluation information can be used for a variety of in-service staff development activities ranging from workshops to research projects and self-assessment. The evaluation reports should help counselors, teachers, and other guidance personnel better understand student needs, the relative effectiveness of guidance activities, and promising new approaches to fulfilling their functions in the educational setting. The following suggestions may be helpful in planning staff development activities using evaluation information:

---

1. Orientation of new staff members to the organization and functioning of the guidance program.
2. Feedback sessions to develop a fuller staff understanding of the major strengths and weaknesses of the guidance program. Evaluation without feedback to those involved cannot be justified. Reinforcement of successes can be motivating for the staff, whereas the work in doing an evaluation without feedback can be demoralizing.
3. Staff workshops to examine the nature and implications of student needs for the guidance program. This type of activity can be helpful particularly in developing staff understanding of the interrelatedness of student needs and the corresponding contributions of the various guidance activities in responding to those needs. Teachers, counselors, and other school staff may be so involved in their special interests that they lost sight of the necessity to correlate their activities with those of the rest of the staff. This is particularly true where some staff focus on crisis needs and others on developmental needs.
4. Staff-centered program development workshops. Evaluation reports can provide the input to stimulate staff efforts in self-examination and program improvement. The examination of the relevance of objectives and of process materials and activities can lead to staff interest in creating and initiating program materials and activities supported by the evaluation evidence.

Program Decisions

Evaluation information provide the basis for making program decisions. These decisions vary from the broad general issues, such as what activities should be provided, to very specific problems, such as what techniques are most effective in facilitating career decision making among minority group tenth-grade students. The value of evaluation results to the program decision-making are related to the evaluation questions asked. Here are some to consider: What are the priority student needs that can be served by the guidance program? Which guidance outcome should receive the highest priority in program planning? What is the relative effectiveness of different activities or techniques in achieving specific outcomes? What is the response of the guidance staff and the students to the different guidance procedures and techniques? What are the possible side effects and the procedures and techniques that are not directly related to the outcomes? What crucial professional program questions have not been answered by the evaluation information and where is there need for further investigation?

Administrative Decisions

Evaluation reports are of value to those responsible for decisions regarding organizational patterns, personnel assignments, and resources management. Most evaluations do not provide direct answers to typical administrative questions but rather provide information from which inferences can be drawn for administrative decision making. Evaluation information may provide input relevant to such administrative questions as, What type of relationship among the guidance personnel produces harmonious and efficient operations, and the expected outcomes? Where are the personnel strengths and weaknesses in the guidance department? What are the characteristics of the most effective staff members? What was the cost of the outcomes observed? Was the outcome of sufficient significance to justify the cost? Where can shifts in personnel or other resources produce outcomes more effectively? Where should priorities be placed in the allocation of resources for the guidance program?

It is important that your staff be involved in the interpretation of the evaluation results for administrative purposes. Your professional explanations of evaluation results will improve the validity of any inferences that are drawn. Your explanations will contribute to the accurate interpretation of evaluation results.

A STUDENT COMPETENCY REPORTING SYSTEM

So far we have discussed traditional approaches to evaluating a comprehensive, developmental guidance program. These approaches play a vital role in an overall evaluation plan, and because they do, they are discussed in detail in most books on the organization and administration of guidance programs. In addition, however, we recommend another approach, an approach that features a student competency reporting system. Keep in mind we did not say "instead of"; we said "in addition to." Process and product evaluation are mandatory in an overall evaluation plan. We feel, however, that an important dimension of guidance program evaluation has been missing: How can students involved in the guidance program participate in evaluation? How can they join with others to monitor their own life career development? Thus we recommend that establishment of a student competency reporting system for the guidance program.

In one way this is not a new idea. Years ago many report cards had an item on it called conduct. Sometimes the grade for conduct was related to academic subjects, but often it represented a subjective rating of how students behaved in school. As educational philosophies changed, conduct changed too. It became industry and attitude. The rating shifted to a judgment of student work habits and how students related with others. Then it became work habits and social growth. Subcategories rated for work habits included "works independently," "follows directions," "listens carefully," "utilizes time appropriately," and "completes daily work." Subcategories rated for the social growth included "gets along well with others," "accepts correction," "practices courtesy," "accepts responsibility," "exhibits appropriate school behavior," and "practices self-control."

Though the use of subcategories was an improvement in communicating what was involved in work habits and social growth, it did not keep pace with what was happening in other curriculum areas. Today continuous progress reporting of students' growth regardless of grade placement is common pactice. Reading levels and mathematics levels are related from year to year in developmental sequence. Health and physical education reporting systems now describe year-by-year growth in students' ability to perform specified physical activities. Where there was once a single-sheet report card for all instruction areas, there is a trend now toward having each instruction area have a report card to be included in a report card packet. The newer reporting systems reflect increasing movement toward individual, developmental instruction programs.

We recommend that you consider developing a student competency reporting system for your guidance program that complements systems now being used in some areas of the instruction program. This system would not take the place of process or product evaluation we discussed previously. They are an important part of your evaluation plan. In fact, the student competency reporting system is a type of product evaluation. But it yields much more than traditional approaches to product evaluation because it becomes a common joining point for students, parents, and staff to take responsibility for the outcomes of the guidance program.

The Basis for the System

The basis for the student competency reporting system are the student competencies you have chosen to assist students to acquire as they are involved in your K–12 guidance program. You will recall that in the guidance curriculum presented in Chapter 3, there were fifteen goals grouped in three domains. For each goal there were thirteen competencies, one for each grade level K–12. A list of these goals and competencies for grades K–12 appears in the Appendix.

To illustrate the basis for a reporting system more specifically, we selected Goal A from the Self-knowledge and Interpersonal Skills Domain. Goal A contains thirteen competencies, one for each grade level. For this example we will use the competency for grade 3 and the competency for grade 12.

Goal A: Students will develop and incorporate an understanding of the unique personal characteristics and abilities of themselves and others.

Grade 3 competency: Students will describe themselves accurately to someone who does not know them.

Grade 12 competency: Students will appreciate their uniqueness and encourage that uniqueness.

The next step is to select performance indicators that will show whether or not students have acquired these competencies.

Also needed are performance levels. Example performance indicators and suggested performance levels for the grade-3 and grade-12 competencies are as follows:

Grade 3

Performance indicators	Suggested performance level
Students are able to	Students are able to
1. describe the physical and personality characteristics that they would like others to know about them.	1. describe six (three physical and three personality) characteristics of themselves.
2. recognize any discrepancies in their descriptions and correct them.	2. correct any incorrect descriptions that they gave for #1.
3. describe themselves correctly to someone they are meeting for the first time.	3. describe themselves to one person they do not know well.

Grade 12

Performance indicators	Suggested performance levels
Students are able to	Students are able to
1. describe their uniqueness and why they appreciate that uniqueness.	1. a) describe two ways they are unique. b) describe their appreciation of that uniqueness in respect to benefits for self and others, and the effects upon the feelings of self and others.
2. describe methods they presently use to encourage their uniqueness.	2. describe three methods presently used to encourage uniqueness (for example, time and effort in learning, practicing, evaluating responses, and so on).
3. predict methods they might use in the future to encourage their uniqueness.	3. predict two methods they might use in the future (for example, time and effort in further practice, further learning, further evaluation or responses, and so on).

Suggested Formats for a Reporting System

There are a variety of formats you could use to organize competencies for reporting purposes. We suggest two. The first one is a report card format similar to report cards now being used in some instruction areas. The other is a folder format similar to ones being used in advisor-advisee systems, particularly in schools where individual planning is featured.

Report Card Format

The report card format being suggested is similar to report card formats being used in a number of instruction areas, such as reading and physical education. Specific competencies are listed on a report form with some indication as to whether or not students have attained these competencies or if they still are working on them. In this format the ratings are done by teachers or counselors in consultation with students. The report is then shared with parents, as are reports from other instruction areas. Performance indicators and performance levels are used in the rating process. An example of a report card for the fifteen competencies for grade 3 follows:

Guidance Program
Third Grade Competencies

Your Child Can	Quarters 1	2	3	4
Describe himself or herself accurately				
Describe personal mental health care				
Describe adult responsibilities				
Recognize actions affect others' feelings				
Talk and listen to close friends and those who are not close friends				
Realize study skills are necessary for learning school subjects				

Your Child Can	1	2	3	4
Define consumer; describe how he or she is a consumer				
Recognize that people have varying roles; describe personal roles				
Recognize why work activities are chosen and that choices may change				
Define what "future" means				
Realize that people obtain rewards for their work				
Recognize those accomplishments he or she is proud of				
Describe personal thought processes before making a decision				
Recognize the need to assess possible consequences before making a decision				
Realize that environment influences interests and capabilities				

The Rating Key

√ Student has accomplished this competency

W Student is still working on this competency

☐ This competency does not apply at this time
(Blank)

Folder Format

Another format you may wish to consider is a folder. The folder would belong to students. It would be a vehicle for them to keep a record of such things as their academic progress, extracurricular activities, important conferences, and future education and work plans. Students at Ocean View High School in Huntington Beach, California, use this approach. At Ocean View High School it is called an *advisee folder.*

To help students keep track of the competencies they are acquiring, using a folder format requires the development of competency lists that can be made a part of the folder. We suggest that individual sheets of heavy stock paper be used to list the fifteen competencies for each grade level included in the folder.

The format might look something like this:

Guidance Program
Twelfth Grade Competencies

Students's Initials	Counselor's Initials	I Can
______	______	Appreciate and encourage my uniqueness.
______	______	Analyze my personal skills that have contributed to satisfactory physical and mental health.
______	______	Assess how taking responsibility enhances my life.
______	______	Understand the value of maintaining effective relationships in today's interdependent society.
______	______	Evaluate my current communication skills; continually improve those skills.
______	______	Evaluate ways I presently learn; predict how learning may change.
______	______	Analyze how I as a citizen and consumer help support the economic system.
______	______	Assess the interactive effects of life roles, settings, and events and how they lead to a personal life-style.
______	______	Analyze the effects stereotypes have on career identity.
______	______	Analyze how concerns change as situations and roles change.
______	______	Speculate what my rights and obligations might be as a producer in the future.
______	______	Summarize the importance of understanding attitudes and values and how they affect my life.
______	______	Implement the decision-making process when making a decision.
______	______	Provide examples and evaluate my present ability to generate alternatives, gather information, and assess consequences in the decisions I make.
______	______	Assess my ability to achieve past goals; integrate this knowledge for my future.

Competency rating using this format could be done jointly by a student and a counselor or an advisor in an advisor-advisee system. The rating would be done by their initialing a competency on the form when they felt it had been acquired. The performance indicators and performance levels for each competency would be used as a basis for making a judgment.

Suggested Uses of a Student Competency Reporting System

Numerous books have been published in the past few years that provided structured approaches to life career planning. Many of these books contained forms to fill out and exercises to complete. While the exercises in these books have been helpful to people, they were often limited by time. Once the activities were completed by an individual, there was little or no provision for follow-through. Some books did provide follow-through exercises, but those, too, were limited.

Recently, however, attention has been given to the follow through issue for special populations by the use of individual education and career development plans and programs. In special education there are *individualized education programs* (IEPs) while in rehabilitation work they are called *individual written rehabilitation plans* (IWRPs) and in Comprehensive Employment Training Programs they are called employability development plans (EDPs).

What is needed in schools is a similar vehicle but for all students. What is needed for each student is an *individual life career plan* (ILCP) in folder form. The plan could be both an instrument and a process that students could use to create and monitor their own development. As an instrument the plan could provide a way for students to gather, analyze, and synthesize information about self and their environment. As a process the plan could become a vehicle through which such information is incorporated into short-range and long-range goal setting, decision making, and planning activities. As a process the plan could become a pathway, a guide that students could follow. It would not be a track that would be plotted and followed routinely. Rather, it would be an outline or plan for a quest.

Central to an individual-life-career-plan folder would be the student competency reporting system. The reporting system would provide a mechanism for students to monitor and record their progress in competency acquisition. Lists of competencies from instruction areas could be added easily to the lists of competencies that result from the guidance program.

To implement this idea, we recommend that individual-life-career-planning folders be established for all students, beginning in th elementary school years. The folders would be the property of the students, although they would be maintained in the guidance office. They would be available to students to use in various guidance and instruction activities related to goal setting, decision making, and planning and would be theirs to take with them when they graduated, transferred, or left school. We further recommend that, concurrently, a report card be developed to share with parents the progress of their children in guidance competency acquisition. The use of a report card would be appropriate particularly during the elementary school years and possibly during the middle school or junior high years.

When students leave school, they would take their individual-life-career-plan folder with them. Whether they would go to work or continue more education, the folder and the accompanying competency lists would be available for additional goal setting, decision making, and planning activities. Information in the folder would assist them in a variety of job-seeking and job-keeping activities, including filling out application forms, writing résumés, developing curriculum vitae, or preparing for job advancement. As new experiences are acquired, they would be analyzed and added to the appropriate sections of the folder. Thus the individual-life-career-plan folder with accompanying competency lists could become an ongoing goal-setting and planning vehicle for individuals as long as they would wish to use it.

REFERENCE

Upton, A. L., B. Lowrey, A. M. Mitchell, B. Varenhorst, and J. Benvenuti, *A Planning Model for Developing a Career Guidance Curriculum.* Fullerton, Calif.: California Personnel and Guidance Association, 1978.

Appendix

STUDENT COMPETENCIES BY DOMAINS AND GOALS

I. Self-knowledge and Interpersonal Skills

A. Students will develop and incorporate an understanding of the unique personal characteristics and abilities of themselves and others.

1. Students will be aware of the unique personal characteristics of themselves and others.
Students will

a. describe their appearance and their favorite activities. (Kindergarten)

b. recognize special or unusual characteristics about themselves. (first grade)

c. recognize special or unusual characteristics about others. (second grade)

d. describe themselves accurately to someone who does not know them. (third grade)

2. Students will demonstrate an understanding of the importance of unique personal characteristics and abilities in themselves and others.
Students will

a. analyze how people are different and how they have different skills and abilities. (fourth grade)

b. specify those personal characteristics and abilities that they value. (fifth grade)

c. analyze how characteristics and abilities change and how they can be expanded. (sixth grade)

d. compare their characteristics and abilities with those of others and will accept the differences. (seventh grade)

e. describe their present skills and predict future skills. (eighth grade)

f. value their unique characteristics and abilities. (ninth grade)

g. analyze how characteristics and abilities develop. (tenth grade)

3. Students will appreciate and encourage the unique personal characteristics and abilities of themselves and others.
Students will

a. specify which characteristics and abilities they appreciate most in themselves and others. (eleventh grade)

b. appreciate their uniqueness and encourage that uniqueness. (twelfth grade)

B. Students will develop and incorporate personal skills that will lead to satisfactory physical and mental health.

1. Students will be aware of personal skills necessary for satisfactory physical and mental health.
Students will

a. describe ways they care for themselves. (Kindergarten)

b. describe how exercise and nutrition affect their mental health. (first grade)

c. describe how they care for their physical health. (second grade)

d. describe how they care for their mental health. (third grade)

e. recognize that they are important to themselves and others. (fourth grade)

f. determine those situations that produce unhappy or angry feelings and how they deal with those feelings. (fifth grade)

g. understand what "stress" means and describe methods of relaxation for handling stress. (sixth grade)

2. Students will demonstrate personal skills that will lead to satisfactory physical and mental health.
Students will

a. distinguish between things helpful and harmful to physical health. (seventh grade)

b. distinguish between things helpful and harmful to mental health. (eighth grade)

c. predict methods they may use in caring for medical emergencies. (ninth grade)

3. Students will demonstrate satisfactory physical and mental health.
Students will

a. effectively reduce their stress during tension-producing situations. (tenth grade)

b. continually evaluate the effects their leisure-time activities have on their physical and mental health. (eleventh grade)

c. analyze their own personal skills that have contributed to satifactory physical and mental health. (twelfth grade)

C. Students will develop and incorporate an ability to assume responsibility for themselves and to manage their environment.

1. Students will be aware of their responsibilities in their environment.
Students will

a. describe areas where they are self-sufficient. (Kindergarten)

b. describe responsibilities they have in their environment. (first grade)

c. give such examples of their environment as their address and the way from school to home. (second grade)

d. describe the responsibilities of adults they know. (third grade)

2. Students will understand the importance of assuming responsibility for themselves and for managing their environment.
Students will

a. know their responsibilities and can be trusted to do them. (fourth grade)

b. analyze how growing up requires more self-control. (fifth grade)

c. know their responsibilities and evaluate their effect on others. (sixth grade)

d. compare and contrast the responsibilities of others in their environment. (seventh grade)

e. evaluate how responsibility helps manage their lives. (eighth grade)

f. analyze when they take responsibility for themselves and when they do not. (ninth grade)

3. Students will assume responsibility for themselves and manage their environment.
   Students will
   a. show how they manage their environment. (tenth grade)
   b. assess how those times of avoiding responsibility hinders their ability to manage their environment effectively. (eleventh grade)
   c. assess how taking responsibility enhances their lives. (twelfth grade)

D. Students will develop and incorporate the ability to maintain effective relationships with peers and adults.
   1. Students will be aware of their relationships with peers and adults.
      Students will
      a. describe their work and play relationships with others. (Kindergarten)
      b. describe the process of making a friend. (first grade)
      c. describe the process of making and keeping a friend. (second grade)
      d. recognize the actions they take that affect others' feelings. (third grade)
      e. indicate methods that lead to effective cooperation with children and adults. (fourth grade)
      f. describe their relationships with family members. (fifth grade)
   2. Students will demonstrate a growing ability to create and maintain effective relationships with peers and adults.
      Students will
      a. analyze the skills needed to make and keep friendships. (sixth grade)
      b. evaluate ways peers and adults interact. (seventh grade)
      c. analyze effective family relationships, their importance, and how they are formed. (eighth grade)
      d. evaluate the importance of having friendships with peers and adults. (ninth grade)
      e. describe situations where their behaviors affect others' behaviors toward them. (tenth grade)
      f. assess their current social and family relationships and evaluate their effectiveness. (eleventh grade)
   3. Students will maintain effective relationships with peers and adults.

Students will

a. understand the value of maintaining effective relationships throughout life in today's interdependent society. (twelfth grade)

E. Students will develop and incorporate listening and expression skills that allow for involvement with others in problem-solving and helping relationships.

1. Students will be aware of listening and expression skills that allow for involvement with others.
Students will

a. recognize that they listen to and speak with a variety of people. (Kindergarten)

b. describe those methods that enable them to speak so they can be understood by others. (first grade)

c. describe those listening and expression skills that allow them to understand others and others to understand them. (second grade)

2. Students will use listening and expression skills that allow for involvement with others.
Students will

a. listen to and speak with both friends and others that are not close friends. (third grade)

b. evaluate how what they say affects others' actions and how what others say affects their actions. (fourth grade)

c. evaluate ways others listen and express thoughts and feelings to them. (fifth grade)

d. use effective nonverbal communication. (sixth grade)

e. evaluate how listening and talking help to solve problems. (seventh grade)

f. analyze how communications skills improve their relationships with others. (eighth grade)

g. analyze how communications skills contribute toward work within a group. (ninth grade)

3. Students will use listening and expression skills that allow for involvement with others in problem-solving and helping relationships.
Students will

a. use communications skills to help others. (tenth grade)

b. analyze how their communications skills encourage problem solving. (eleventh grade)

c. evaluate their current communication skills and continually improve those skills. (twelfth grade)

II. Life Roles, Settings, and Events

A. Students will develop and incorporate those skills that lead to an effective role as a learner.

1. Students will be aware of themselves as learners.
Students will

a. describe those things they learn at school. (Kindergarten)

b. relate learning experiences at school to situations in the home. (first grade)

c. recognize some benefits of learning. (second grade)

d. realize that certain study skills are necessary for learning each school subject. (third grade)

e. describe the various methods they use to learn in school. (fourth grade)

2. Students will use those skills that lead to an effective role as a learner.
Students will

a. analyze how their basic study skills relate to desired work skills. (fifth grade)

b. analyze how school learning experiences relate to their leisure activities. (sixth grade)

c. predict how they will use knowledge from certain subjects in future life and work experiences. (seventh grade)

d. learn, both in and out of the school setting. (eighth grade)

e. describe personal learning and study skills and explain their importance. (ninth grade)

f. evaluate personal learning and study skills and explain how they can be improved. (tenth grade)

3. Students will be effective in their roles as learners.
Students will

a. predict how their developed learning and study skills can contribute to work habits in the future. (eleventh grade)

b. evaluate ways they presently learn and predict how learning may change in the future. (twelfth grade)

B. Students will develop and incorporate an understanding of the legal and economic principles and practices that lead to responsible daily living.

1. Students will be aware of legal and economic principles and practices.

Students will

a. recognize the town, state, and country in which they reside. (Kindergarten)

b. understand why people use money in our economic system. (first grade)

c. describe those rules they follow in their environment and why those rules are necessary. (second grade)

d. understand what a consumer is and how they are consumers. (third grade)

e. describe how people depend on each other to fulfill their needs. (fourth grade)

f. recognize that a wage earner is required to pay taxes. (fifth grade)

g. describe how the government uses tax money. (sixth grade)

2. Students will demonstrate a growing ability to use legal and economic principles and practices that lead to responsible daily living.
Students will

a. describe the rights and responsibilities they have as citizens of their towns and states. (seventh grade)

b. describe the rights and responsibilities they have as U.S. citizens. (eighth grade)

c. evaluate the purposes of taxes and how they support the government. (ninth grade)

d. evaluate their roles as consumers. (tenth grade)

e. analyze their legal rights and responsibilities as consumers. (eleventh grade)

3. Students will use responsible legal and economic principles and practices in their daily lives.
Students will

a. analyze how they, as citizens and consumers, help to support the economic system. (twelfth grade)

C. Students will develop and incorporate an understanding of the interactive effects of life-styles, life roles, settings, and events.

1. Students will be aware of life-styles, life roles, settings, and events.
Students will

a. describe their daily activities at school. (Kindergarten)

b. realize how they have changed during the past year. (first grade)

c. describe necessary daily activities that are carried out by self and others. (second grade)

d. recognize that people have varying roles and describe their own roles. (third grade)

e. understand what important events affect the lives of self and others. (fourth grade)

f. recognize what a life-style is and what influences their life-styles. (fifth grade)

2. Students will acknowledge the interactive effects of life-styles, life roles, settings, and events in their lives.
Students will

a. analyze ways they have control over themselves and their life-styles. (sixth grade)

b. evaluate their feelings in a variety of settings. (seventh grade)

c. predict their feelings in a variety of potential settings. (eighth grade)

d. analyze how life roles, settings, and events determine preferred life-styles. (ninth grade)

e. compare how life-styles differ depending on life roles, settings, and events. (tenth grade)

3. Students will understand the interactive effects of life-styles, life roles, settings, and events.
Students will

a. determine how life roles, settings, and events have influenced their present life-styles. (eleventh grade)

b. assess the interactive effects of life roles, settings, and events and how these lead to a preferred life-style. (twelfth grade)

D. Students will develop and incorporate an understanding of stereotypes and how stereotypes affect career identity.

1. Students will be aware of stereotypes and some of their effects.
Students will

a. mentally project adults into work activities other than those they do presently. (Kindergarten)

b. recognize how peers differ from themselves. (first grade)

c. distinguish which work activities in their environment are done by certain people. (second grade)

d. recognize why people choose certain work activities and that those choices may change. (third grade)

e. define the meaning of "stereotypes" and indicate how stereotypes affect them. (fourth grade)

f. describe stereotypes that correspond with certain jobs. (fifth grade)

2. Students will demonstrate a growing ability to understand stereotypes and how stereotypes affect career identity.
Students will
   a. predict how stereotypes might affect them in work activities. (sixth grade)
   b. describe occupations that have stereotypes existing for them and will analyze how those stereotypes are reinforced. (seventh grade)
   c. evaluate the ways in which certain groups (men, women, minorities, and so on) are stereotyped. (eighth grade)
   d. analyze stereotypes that exist for them and how those stereotypes limit their choices. (ninth grade)
   e. analyze stereotypes held by others and how those stereotypes can limit choices. (tenth grade)

3. Students will understand stereotypes in their lives and environment and how stereotypes affect career identity.
Students will
   a. evaluate their stereotypes and explain those they have changed. (eleventh grade)
   b. analyze the effect stereotypes have on career identity. (twelfth grade)

E. Students will develop and incorporate the ability to express futuristic concerns and the ability to imagine themselves in these situations.

1. Students will be aware of the future and what situations might occur in the future.
Students will
   a. describe situations that are going to happen in the future. (Kindergarten)
   b. describe situations desired for the future and when they would like those situations to happen. (first grade)
   c. recognize what they would like to accomplish when they are three years older. (second grade)
   d. define what "future" means. (third grade)

2. Students will demonstrate a growing ability to express futuristic concerns and to imagine themselves in such situations.
Students will
   a. imagine what their lives must be like in the future. (fourth grade)
   b. imagine what the world will be like in twenty years. (fifth grade)

c. predict what they will be like in twenty years. (sixth grade)

d. predict ways in which some present careers may be different in the future. (seventh grade)

e. predict how they may have to change to fit into a career in the future (eighth grade)

f. analyze how choices they are making now will affect their lives in the future. (ninth grade)

g. predict some of the concerns they will have as they get older. (tenth grade)

3. Students will express futuristic concerns and will imagine themselves in these situations.
Students will

a. evaluate the need for flexibility in their roles and in their choices. (eleventh grade)

b. analyze how concerns change as situations and roles change. (twelfth grade)

III. Life Career Planning

A. Students will develop and incorporate an understanding of producer rights and responsibilities.

1. Students will be aware of what a producer is and that producers have rights and responsibilities.
Students will

a. describe the work activities of family members. (Kindergarten)

b. describe different work-activities and their importance. (first grade)

c. define "work" and recognize that all people work. (second grade)

d. realize that people obtain rewards for their work. (third grade)

e. recognize that a producer can have many different roles. (fourth grade)

f. recognize how they depend on different producers. (fifth grade)

2. Students will demonstrate an understanding of what a producer is and producer rights and responsibilities.
Students will

a. demonstrate steps they follow in producing a product or task they take pride in. (sixth grade)

b. show appreciation when others successfully complete a difficult task. (seventh grade)

c. analyze the relationship between interests and producer satisfaction. (eighth grade)

d. analyze how producers may have to cooperate with each other to accomplish a large or difficult task. (ninth grade)

e. evaluate the importance of having laws and contracts to protect producers. (tenth grade)

3. Students will understand and exemplify producer rights and responsibilities.
   Students will

   a. specify their rights and responsibilities as producers. (eleventh grade)

   b. speculate what their rights and obligations might be as producers in the future. (twelfth grade)

B. Students will develop and incorporate an understanding of how attitudes and values affect decisions, actions, and life-styles.

1. Students will be aware of attitudes and values and their effects.
   Students will

   a. describe people and activities they enjoy. (Kindergarten)

   b. describe actions of others that they do not appreciate. (first grade)

   c. describe those things they have learned that aid in making choices. (second grade)

   d. recognize those accomplishments they are proud of. (third grade)

   e. define "attitudes" and "beliefs" and describe the effects attitudes and beliefs have on decisions. (fourth grade)

   f. define "values" and describe their own values. (fifth grade)

2. Students will demonstrate a growing understanding of how attitudes and values affect decisions, actions, and life-styles.
   Students will

   a. analyze how their attitudes and values influence what they do. (sixth grade)

   b. compare and contrast others' values. (seventh grade)

   c. predict how their values will influence their life-styles. (eighth grade)

   d. describe and prioritize their values. (ninth grade)

   e. describe decisions they have made that were based on their attitudes and values. (tenth grade)

3. Students will understand how attitudes and values affect decisions, actions, and life-styles.
   Students will
   a. analyze how values affect their decisions, actions, and life-styles. (eleventh grade)
   b. summarize the importance of understanding their attitudes and values and how those attitudes and values affect their lives. (twelfth grade)

C. Students will develop and incorporate an understanding of the decision-making process and how the decisions they make are influenced by previous decisions made by themselves and others.
   1. Students will be aware of decisions and the decision-making process.
      Students will
      a. describe choices they make. (Kindergarten)
      b. describe decisions they make by themselves. (first grade)
      c. recognize why some choices are made for them; they can accept those choices and make their own decisions when appropriate. (second grade)
      d. describe their thought processes before a decision is made. (third grade)
      e. describe why they might want to change a decision and recognize when it is or is not possible to make that change. (fourth grade)
      f. describe the decision-making process. (fifth grade)
      g. recognize how school decisions influence them. (sixth grade)
   2. Students will understand the decision-making process and those factors that influence the decisions they make.
      Students will
      a. provide examples of how past decisions they have made influence their present actions. (seventh grade)
      b. analyze how past decisions made by their families influence their present decisions. (eighth grade)
      c. evaluate the influence that past legal decisions have on their present decisions. (ninth grade)
      d. analyze the decision-making process used by others. (tenth grade)
   3. Students will effectively use the decision-making process and understand how the decisions they make are influenced by previous decisions made by themselves and others.

Students will

a. identify decisions they have made and analyze how those decisions will affect their future decisions. (eleventh grade)

b. implement the decision-making process when making a decision. (twelfth grade)

D. Students will develop and incorporate the ability to generate decision-making alternatives, gather necessary information, and assess the risks and consequences of alternatives.

1. Students will be aware of method of generating decision-making alternatives, gather necessary information, and assessing the risks and consequences of alternatives.
Students will

a. realize the difficulty of making choices between two desirable alternatives. (Kindergarten)

b. recognize which decisions are difficult for them. (first grade)

c. realize that they go through a decision-making process each time they make a choice. (second grade)

d. recognize that they are able to assess possible consequences of a decision before actually making the choice. (third grade)

2. Students will demonstrate a growing ability to generate decision-making alternatives, gather necessary information, and assess the risks and consequences of alternatives.
Students will

a. generate alternatives to a specific decision. (fourth grade)

b. evaluate some of the risks involved in choosing one alternative over another. (fifth grade)

c. consider the results of various alternatives and then make their choice. (sixth grade)

d. provide examples of some consequences of a decision. (seventh grade)

e. demonstrate how gaining more information increases their alternatives. (eighth grade)

f. analyze the importance of generating alternatives and assessing the consequences of each before making a decision. (ninth grade)

g. distinguish between alternatives that involve varying degrees of risks. (tenth grade)

h. analyze the consequences of decisions that others make. (eleventh grade)

3. Students will generate decision-making alternatives, gather necessary information, and assess the risks and consequences of alternatives.
Students will

a. provide examples and evaluate their present ability to generate alternatives, gather information, and assess the consequences in the decisions they make. (twelfth grade)

E. Students will develop and incorporate skill in clarifying values, expanding interests and capabilities, and evaluating progress toward goals.

1. Students will be aware of values, interests and capabilities, and methods of evaluation.
Students will

a. describe growing capabilities. (Kindergarten)

b. identify those capabilities they wish to develop. (first grade)

c. recognize activities that interest them and those that do not. (second grade)

d. realize that environment influences interests and capabilities. (third grade)

e. recognize different methods of evaluating task progress. (fourth grade)

f. describe the meaning of "value" and how values contribute toward goal decisions. (fifth grade)

2. Students will gain skill in clarifying values, expanding interests and capabilities, and evaluating progress toward goals.
Students will

a. predict five goals (based on their interests and capabilities) they would like to achieve within five years. (sixth grade)

b. analyze various methods of evaluating their progress toward a goal. (seventh grade)

c. contrast goals they desire to complete with goals they expect to complete. (eighth grade)

d. define their unique values, interests, and capabilities. (ninth grade)

e. evaluate the importance of setting realistic goals and striving toward them. (tenth grade)

3. Students will clarify their values, expand their interests and capabilities, and evaluate their progress toward goals.

Students will

a. analyze how their values, interests, and capabilities have changed and are changing. (eleventh grade)

b. assess their ability to achieve past goals and integrate this knowledge for the future. (twelfth grade)

STUDENT COMPETENCIES BY GRADE LEVEL, K-12

Domain I: Self-knowledge and Interpersonal Skills

Goal A: Students will develop and incorporate an understanding of the unique personal characteristics and abilities of themselves and others.

K. I can tell what I look like and some things I like to do.
1. I can tell something special about myself.
2. I can tell something special about other people I know.
3. I can describe myself to someone who doesn't know me.
4. I can tell how people are different and that they have different skills and abilities.
5. I can tell how my special characteristics and abilities are important to me.
6. I can tell how my characteristics and abilities change and how they can be expanded.
7. I can compare the characteristics and abilities of others I know with my own and accept the differences.
8. I can list the skills I already possess and those I hope to develop in the future.
9. I can discuss the value of understanding my unique characteristics and abilities.
10. I can describe and analyze how an individual's characteristics and abilities develop.
11. I can explain which characteristics and abilities I appreciate most in myself and others.
12. I can compare my characteristics and abilities with those of others and appreciate and encourage my uniqueness.

Domain I: Self-knowledge and Interpersonal Skills

Goal B: Students will develop and incorporate personal skills that will lead to satisfactory physical and mental health.

K. I can tell some ways I take care of myself.
1. I can tell how exercise and eating habits make a difference in how I think and act.
2. I can tell how I keep by body healthy.
3. I can tell how I keep myself feeling happy.
4. I can tell why I am an important person to myself and others.
5. I can tell what happens to make me unhappy or angry and how I deal with those feelings.
6. I can define the word *stress* and tell some ways I relax when I feel stress.
7. I can distinguish between things helpful and harmful to physical health.
8. I can distinguish between things helpful and harmful to mental health.
9. I can list ways to care for medical emergencies.
10. I can list three high-stress situations for me and tell how I effectively reduce my stress.
11. I can evaluate ways I spend my leisure time and how it affects my physical and mental health.
12. I can analyze my health and how I must integrate aspects of physical and mental health for me to be totally happy.

Domain I: Self-knowledge and Interpersonal Skills

Goal C: Students will develop and incorporate an ability to assume responsibility for themselves and to manage their environment.

K. I can tell some things I can do for myself.
1. I can tell some jobs I do at home and at school.
2. I can tell my address and how to get from school to my home.
3. I can tell about some of the responsibilities of adults I know.
4. I can tell what my jobs are, and I can be trusted to do them.

5. I can tell how growing up requires more self-control.
6. I can tell about some of my responsibilities and the effect they have on others.
7. I can list the responsibilities people have at school and tell why some people have more or different responsibilities than others.
8. I can explain how responsibility can help me manage my life.
9. I can describe when I take responsibility for myself and when I do not.
10. I can provide examples of the ways I manage my environment.
11. I can explain how avoiding responsibility for my actions hinders my ability to manage my environment.
12. I can assess how taking responsibility for myself enhances my life.

Domain I: Self-knowledge and Interpersonal Skills

Goal D: Students will develop and incorporate the ability to maintain effective relationships with peers and adults.

K. I can tell how I work and play with others.
1. I can tell how I make a friend.
2. I can tell how I make and keep a friend.
3. I can tell the things I do to make others feel good or bad.
4. I can tell how I cooperate with children and adults both at school and at home.
5. I can tell about my relationships with those in my family.
6. I can discuss skills that help me make and keep friends.
7. I can describe ways peers and adults interact.
8. I can tell why good family relations are important and how they are formed.
9. I can explain the importance of friendship with peers and adults.
10. I can provide examples of ways my behaviors affect others' behaviors toward me.
11. I can assess my current social and family relationships and evaluate their effectiveness.
12. I can discuss the value of maintaining effective relationships throughout life in today's interdependent society.

Domain I: Self-knowledge and Interpersonal Skills

Goal E: Students will develop and incorporate listening and expression skills that allow for involvement with others in problem-solving and helping relationships.

K. I can listen and talk with different people.
1. I can talk so others can understand what I am saying.
2. I can listen and speak so that I can understand others and they can understand me.
3. I can talk and listen to those I choose as friends and those who are not close friends.
4. I can describe how what I say affects what other people do, and what others say affects what I do.
5. I can list ways that others listen and express their thoughts and feelings to me.
6. I can list ways in which I communicate nonverbally with others.
7. I can explain how listening and talking helps solve problems.
8. I can explain how communication skills can improve my relationship with others.
9. I can explain how communication skills help me work within a group.
10. I can provide examples of situations in which my communication skills helped others.
11. I can summarize ways that communication skills encourage problem solving.
12. I can evaluate my current communication skills and consider how they can be improved.

Domain II: Life Roles, Settings, and Events

Goal A: Students will develop and incorporate those skills that lead to an effective role as a learner.

K. I can tell things I am learning at school each day.
1. I can tell about things I see at home that I learned about at school.
2. I can tell how learning helps me to do things I like.
3. I can list which study skills are needed for each subject.
4. I can describe the different ways I learn in school.
5. I can relate basic study skills in school to work skills.

6. I can tell how school experiences relate to leisure-time activities.
7. I can explain how I will use knowledge gained from certain subjects in my future life and work experiences.
8. I can describe ways I learn, both in school and out of school.
9. I can identify my learning and study skills and explain why they are important.
10. I can evaluate my learning and study skills and explain how I can improve them.
11. I can analyze how adequate learning and study skills can contribute to work habits in the future.
12. I can evaluate the ways I presently learn and imagine how my learning will change in the future.

Domain II: Life Roles, Settings, and Events

Goal B: Students will develop and incorporate an understanding of the legal and economic principles and practices that lead to responsible daily living.

K. I can tell the name of the town, state, and country I live in.
1. I can tell why we use money.
2. I can tell what rules I have at home and at school and why I need to follow them.
3. I can tell what a consumer is and the ways I am a consumer.
4. I can tell how people depend on one another to fulfill their needs.
5. I can describe how workers keep some of the money they are paid and that they have to give some of the money to state and federal governments.
6. I can tell some things that the government does with workers' tax money.
7. I can list several rights and responsibilities that I have as a citizen of my town and state.
8. I can list several rights and responsibilities I have as a U.S. citizen.
9. I can list several reasons why we pay taxes and how they help support our government.
10. I can evaluate my role as a consumer.
11. I can analyze my legal rights and responsibilities as a consumer.
12. I can analyze how I, as a citizen and a consumer, help to support our economic system.

Domain II: Life Roles, Settings, and Events

Goal C: Students will develop and incorporate an understanding of the interactive effects of life-styles, life roles, settings, and events.

K. I can tell about different kinds of things I do at school each day.
1. I can tell how I have changed in the past year.
2. I can tell some jobs that people have to do each day and why they must be done.
3. I can tell that people have different roles and I can describe my roles.
4. I can describe some important events that affect our lives.
5. I can describe what a life-style is and what influences my life-style.
6. I can list ways that I have control over myself and my life-style.
7. I can describe when I am comfortable or uncomfortable in a variety of environments.
8. I can imagine myself in different environments and I can recognize how I might feel in each one.
9. I can explain how life roles, settings, and events determine preferred life-styles.
10. I can compare how life-styles differ depending on life roles, settings, and events.
11. I can discuss how life roles, settings, and events have influenced my present life-style.
12. I can assess the interactive effects of life roles, settings, and events and how this leads to a preferred life-style.

Domain II: Life Roles, Settings, and Events

Goal D: Students will develop and incorporate an understanding of stereotypes and how stereotypes affect career identity.

K. I can imagine grown-ups doing a job other than the one they do now.
1. I can tell how other children are different from me.
2. I can tell which jobs are done by certain people (at home and at school).
3. I can tell why people choose certain jobs and that they may change them from time to time.

4. I can tell what "stereotypes" are and how they affect me.
5. I can tell about some of the stereotypes that go along with certain jobs.
6. I can describe how stereotypes could affect me in a job.
7. I can list some occupations that have stereotypes existing for them and explain how they are reinforced.
8. I can explain ways in which certain groups (men, women, minorities, and so on) are stereotyped.
9. I can identify stereotypes that exist for me and how these limit my choices.
10. I can recognize stereotypes held by others and how these can limit choices.
11. I can evaluate my stereotypes and explain those I have changed.
12. I can analyze the effect stereotypes have on career identity.

Domain II: Life Roles, Settings, and Events

Goal E: Students will develop and incorporate the ability to express futuristic concerns and the ability to imagine themselves in these situations.

K. I can tell some things that will happen later (for example, tomorrow, next week, next summer).

1. I can tell some things I would like to happen later on and tell when I would like them to happen.
2. I can tell some things I think I would like to do when I am three years older.
3. I can tell what the word *future* means.
4. I can imagine what my life might be like in the future (for example, what my home might look like, what kinds of things I might do, what I might look like).
5. I can imagine different possibilities of what the world will be like in twenty years.
6. I can tell what I would like to be like in twenty years.
7. I can list ways in which some careers now may be different in the future.
8. I can describe how I might have to change to fit into a career in the future.
9. I can describe how choices I am making now will affect my life in the future.

10. I can predict some of the concerns I will have as I get older.
11. I can explain the need for flexibility in my roles and in my choices.
12. I can analyze how concerns change as situations and roles change.

Domain III: Life Career Planning

Goal A: Students will develop and incorporate an understanding of producer rights and responsibilities.

K. I can tell about the different kinds of work each member of my family does.
1. I can tell about some different jobs that people do and why those jobs are important.
2. I can tell what "work" means and I can show that all people work.
3. I can describe some of the rewards people get for their work.
4. I can tell how a producer (worker) can have many different roles.
5. I can describe how we depend on different producers.
6. I can describe the steps that I follow in producing something I can be proud of.
7. I can show appreciation to someone when they successfully finish a difficult task.
8. I can explain the relationship between interests and producer satisfaction.
9. I can describe how producers may have to cooperate with each other to accomplish a large or difficult task.
10. I can evaluate the importance of having laws and contracts to protect producers.
11. I can describe my rights and responsibilities as a producer.
12. I can speculate what my rights and obligations might be as a producer in the future.

Domain III: Life Career Planning

Goal B: Students will develop and incorporate an understanding of how attitudes and values affect decisions, actions, and life-styles.

K. I can tell about people I like and about things I like to do.

1. I can tell some things that I don't like other children to do.
2. I can tell some things that I learn at home and at school that help me make choices.
3. I can tell some things I have done that I am proud of.
4. I can tell what attitudes and beliefs are and how they affect my decisions.
5. I can tell what a value is and describe my own values.
6. I can tell how my attitudes and values influence the things I do.
7. I can list what I think other people's values are.
8. I can show how my values will influence my life-style.
9. I can identify and prioritize my values.
10. I can provide examples of decisions I have made based on my attitudes and values.
11. I can explain and analyze how values affect my decisions, actions, and life-styles.
12. I can summarize the importance of understanding my attitudes and values and how they affect my life.

Domain III: Life Career Planning

Goal C: Students will develop and incorporate an understanding of the decision-making process and how the decisions they make are influenced by previous decisions made by themselves and others.

K. I can tell about choices I make.
1. I can tell which choices I make on my own.
2. I can tell why some choices are made for me; I can accept those choices and make my own choices when I can.
3. I can tell what I think about before I make a decision.
4. I can tell why I might want to change a decision and when it is or is not possible to change a decision.
5. I can tell the steps it takes to make a decision.
6. I can list ways in which school decisions influence me.
7. I can give an example of how something I decided in the past can influence what I do today.
8. I can analyze how past decisions my family has made influence the ones I make now.

9. I can evaluate the influence that past legal decisions have on the decisions I presently make.
10. I can analyze the decision-making process used by others.
11. I can identify decisions I have made and analyze how they will affect my future decisions.
12. I can use the decision-making process when making a decision.

Domain III: Life Career Planning

Goal D: Students will develop and incorporate the ability to generate decision-making alternatives, gather necessary information, and assess the risks and consequences of alternatives.

K. I can make choices between two things I would like to have.
1. I can tell which choices are hard for me to make.
2. I can tell what I think about when I make a choice.
3. I can find out what might happen before I make a choice.
4. I can list alternatives to a specific decision.
5. I can explain some of the risks involved in choosing one alternative over another.
6. I can describe how I think about the results of different actions and then choose what I will do.
7. I can list some consequences of a decision.
8. I can explain how gaining more information increases my alternatives.
9. I can explain the importance of generating alternatives and assessing the consequence of each before making a decision.
10. I can distinguish between alternatives that involve varying degrees of risks.
11. I can analyze the consequences of decisions others make.
12. I can provide examples and evaluate my present ability to generate alternatives, gather information, and assess the consequences in the decisions I make.

Domain III: Life Career Planning

Goal E: Students will develop and incorporate skill in clarifying values, expanding intersts and capabilities, and evaluating progress toward goals.

K. I can name new things I've learned to do that I couldn't do before.
1. I can name some things I want to learn how to do.
2. I can tell some things I like to do and some things I don't like to do.
3. I can tell how my family and school influence what I am interested in and what I can do.
4. I can name different ways of deciding if a job or project is going well or not.
5. I can describe what a "value" is and how it helps in deciding on a goal.
6. I can list five goals I would like to achieve within five years.
7. I can list several ways of evaluating my progress toward a goal.
8. I can explain the difference between a goal I would like to complete and one I expect to complete.
9. I can describe my unique values, interests, and capabilities.
10. I can evaluate the importance of setting realistic goals and working toward them.
11. I can explain how my values, interests, and capabilities have changed and are changing.
12. I can assess my ability to achieve past goals and integrate this knowledge for the future.

Author Index

Subject Index